WASTEWATER AND SHALE FORMATION DEVELOPMENT

Risks, Mitigation, and Regulation

WASTEWATER AND SHALE FORMATION DEVELOPMENT
Risks, Mitigation, and Regulation

Edited by
Sheila Olmstead, PhD

Apple Academic Press Inc. | Apple Academic Press Inc.
3333 Mistwell Crescent | 9 Spinnaker Way
Oakville, ON L6L 0A2 | Waretown, NJ 08758
Canada | USA

©2015 by Apple Academic Press, Inc.

First issued in paperback 2021

Exclusive worldwide distribution by CRC Press, a member of Taylor & Francis Group

No claim to original U.S. Government works

ISBN 13: 978-1-77463-566-7 (pbk)
ISBN 13: 978-1-77188-160-9 (hbk)

This book contains information obtained from authentic and highly regarded sources. Reprinted material is quoted with permission and sources are indicated. Copyright for individual articles remains with the authors as indicated. A wide variety of references are listed. Reasonable efforts have been made to publish reliable data and information, but the authors, editors, and the publisher cannot assume responsibility for the validity of all materials or the consequences of their use. The authors, editors, and the publisher have attempted to trace the copyright holders of all material reproduced in this publication and apologize to copyright holders if permission to publish in this form has not been obtained. If any copyright material has not been acknowledged, please write and let us know so we may rectify in any future reprint.

Trademark Notice: Registered trademark of products or corporate names are used only for explanation and identification without intent to infringe.

Library and Archives Canada Cataloguing in Publication

Wastewater and shale formation development : risks, mitigation, and regulation / edited by Sheila Olmstead, PhD.

Includes bibliographical references and index.
ISBN 978-1-77188-160-9 (bound)
1. Hydraulic fracturing--Environmental aspects. 2. Hydraulic fracturing--Law and legislation. 3. Shale gas industry--Environmental aspects. 4. Shale gas industry--Waste minimization. 5. Shale gas industry--Water-supply. 6. Water--Pollution--Prevention. I. Olmstead, Sheila M., author, editor

TD195.G3W28 2015   622'.3381   C2015-902717-9

Library of Congress Cataloging-in-Publication Data

Wastewater and shale formation development : risks, mitigation, and regulation / Sheila Olmstead, PhD, editor.

pages cm
Includes bibliographical references and index.
ISBN 978-1-77188-160-9 (alk. paper)
1. Oil pollution of groundwater. 2. Oil pollution of water. 3. Hydraulic fracturing--Environmental aspects. 4. Water--Pollution. I. Olmstead, Sheila M., editor.

TD427.P4W377 2015   628.1'6833--dc23   2015014294

Apple Academic Press also publishes its books in a variety of electronic formats. Some content that appears in print may not be available in electronic format. For information about Apple Academic Press products, visit our website at **www.appleacademicpress.com** and the CRC Press website at **www.crc-press.com**

About the Editor

SHEILA OLMSTEAD, PhD

Sheila Olmstead, PhD, is an Associate Professor of Public Affairs at the LBJ School. Before joining the LBJ School, Dr. Olmstead was a Fellow (2010-2013) and Senior Fellow (2013) at Resources for the Future in Washington, DC, as well as an Associate Professor (2007–2010) and Assistant Professor (2002–2007) of Environmental Economics at the Yale University School of Forestry and Environmental Studies, where she was the recipient of three teaching awards. Olmstead is an environmental economist whose current research projects examine the environmental externalities associated with shale gas development in the United States, regulatory avoidance under the U.S. Safe Drinking Water Act, the influence of federal fire suppression policy on land development in the American West, and free-riding in dam placement and water withdrawals in transboundary river basins. She has worked extensively on the economics of water resource management, focusing on water demand estimation, water conservation policy, and access to drinking water services among low-income communities. Climate and energy policy are additional topics of her research, especially with regard to the application of market-based environmental policy instruments. Her research has been published in leading journals such as the *Journal of Economic Perspectives, Proceedings of the National Academy of Sciences, Journal of Business and Economic Statistics, Journal of Environmental Economics and Management, Journal of Urban Economics*, and *Water Resources Research*. With Nathaniel Keohane, she is the author of the 2007 book *Markets and the Environment*. Her research has been funded by the National Science Foundation, U.S. Environmental Protection Agency, U.S. Department of the Interior, World Bank, Alfred P. Sloan Foundation, U.S. Department of Energy, and National Oceanic and Atmospheric Administration. Dr. Olmstead is a former member of the Board of Directors of the Association of Environmental and Resource Economists, and a member of the Advisory Board of the International Water Resource Economics Consortium.

Contents

Part III: The Quest for Mitigation

Part IV: Fracking Wastewater Regulations

Acknowledgment and How to Cite

The editor and publisher thank each of the authors who contributed to this book. The chapters in this book were previously published elsewhere in various formats. To cite the work contained in this book and to view the individual permissions, please refer to the citation at the beginning of each chapter. Each chapter was read individually and carefully selected by the editor; the result is a book that provides a nuanced look at the wastewater issues associated with shale gas development. The chapters included are divided into four sections, which describe the following topics:

- In Part 1, chapters 1, 2, and 3 were selected for their focus on water use and the fate of injected water. The articles consider two very different shale plays (the Marcellus in chapters 1 and 3, and the Barnett, in chapter 2), with completely different water sourcing and wastewater disposal options, which offers the reader a more comprehensive summary of the state-of-the-science.
- In Part 2, which focuses on the environmental impact of hydrofacturing wastewater, chapters 4 and 5 were selected to give research-based analysis from objective sources on the impact of fracking wastewater on surface and ground water. Chapter 6 defines the problem of naturally occurring radioactive material in shale wastewater, which will be discussed further in Part 3's discussion of mitigation techniques.
- In Part 3, chapters 7 and 8 provide research on methodologies and technologies that offer solutions to the problems of fluid recovery in terms of flowback treatment, as well as treatments to deal with radioactivity in shale wastewater.
- In Part 4, chapters 9 and 10 discuss the legal issues and their connections to scientific research, as well as their implications for future environmental and economic consequences. Chapter 10 provides important information on U.S. states' individual regulations, since the states are such key players in oil and gas regulation.
- Chapter 11 is useful as a summary conclusion, restating factors discussed throughout the book at a simplified level and connecting the scientific, environmental, and regulatory issues at stake.

List of Contributors

Mohammed H. Al-Wadei
Department of Public Health, 390 HPER Building, 1914 Andy Holt Avenue, University of Tennessee, Knoxville, TN 37996, USA

Elise Barbot
Department of Civil and Environmental Engineering, University of Pittsburgh, Pittsburgh, Pennsylvania, USA

Valerie J. Brown
Oregon, USA

Terence J. Centner
Department of Agricultural and Applied Economics, College of Agricultural and Environmental Sciences, The University of Georgia, Athens, GA 30602, USA

Jiangang Chen
Department of Public Health, 385 HPER Building, 1914 Andy Holt Avenue, University of Tennessee, Knoxville, TN 37996, USA

Ziyan Chu
Resources for the Future, 1616 P Street NW, Washington, DC 20036, USA

Ruth A. Costley
Bureau of Economic Geology, Jackson School of Geosciences, The University of Texas, Box X, Austin, Texas 78713, USA

Milind Deo
Department of Chemical Engineering, University of Utah, 50 Central Campus Drive, Salt Lake City, UT 84112, USA

Adrian Down
Division of Earth and Ocean Sciences, Nicholas School of the Environment, Duke University, Durham, NC 27708, USA

Khaled Enab
Department of Energy and Mineral Engineering, The Pennsylvania State University, 110 Hosler Building, University Park, PA 16802, USA

Sara Gosman
University of Michigan Law School, 625 South State Street, Ann Arbor, MI 48109, USA

Kelvin B. Gregory
Department of Civil and Environmental Engineering,Carnegie Mellon University, Pittsburgh, PA 15213, USA

Richard W. Hammack
National Energy Technology Laboratory (NETL), Pittsburgh, Pennsylvania PA 15236, USA

Heather Hatzenbuhler
Department of Agricultural and Applied Economics, College of Agricultural and Environmental Sciences, The University of Georgia, Athens, GA 30602, USA

Phillip D. Hays
U.S. Geological Survey, Arkansas Water Science Center, Fayetteville, AR 72701, USA

Robert B. Jackson
School of Earth, Energy and Environmental Sciences, Stanford University, Y2E2 Building, 379B, Stanford, CA 94305, USA

Hongtao Jia
Department of Chemical Engineering, University of Utah, Salt Lake City, UT, USA

Jonathan D. Karr
Duke Environmental Stable Isotope Laboratory, Duke University, Durham, NC 27708, USA

Rebekah C. M. Kennedy
Department of Public Health, 390 HPER Building, 1914 Andy Holt Avenue, University of Tennessee, Knoxville, TN 37996, USA

Timothy M. Kresse
U.S. Geological Survey, Arkansas Water Science Center, Little Rock, AR 72211, USA

Alan J. Krupnick
Resources for the Future, 1616 P Street, NW, Washington, DC 20036, USA

Christian Madu
Department of Energy and Mineral Engineering, The Pennsylvania State University, 110 Hosler Building, University Park, PA 16802, USA

John McLennan
Departent of Chemical Engineering, 50 S. Central Campus Drive, University of Utah, Salt Lake City, UT 84112, USA

Lucija A. Muehlenbachs
Department of Economics, The University of Calgary, 2500 University Drive, NW, Calgary, AB, T2N1N4, Canada; and Resources for the Future, 1616 P Street, NW, Washington, DC 20036, USA

Jean-Philippe Nicot
Bureau of Economic Geology, The University of Texas, Austin, TX 78758, USA

Richard Olawoyin
School of Health Sciences, Oakland University, 2200 N. Suirrel Road, Rochester, MI 48309, USA

Sheila M. Olmstead
LBJ School of Public Affairs, The University of Texas at Austin, PO Box Y, Austin TX 78713, USA; and Resources for the Future, 1616 P Street NW, Washington, DC 20036, USA

Robert C. Reedy
Bureau of Economic Geology, The University of Texas, Austin, TX 78758, USA

Bridget R. Scanlon
Bureau of Economic Geology, Jackson School of Geosciences, The University of Texas, Austin, TX 78758, USA

Jhih-Shyang Shih
Resources for the Future, 1616 P Street, NW, Washington, DC 20036, USA

Paul D. Terry
Department of Public Health, 375 HPER Building, 1914 Andy Holt Avenue, University of Tennessee, Knoxville, TN 37996, USA

Avner Vengosh
Division of Earth and Ocean Sciences, Nicholas School of the Environment, Duke University, Box 27708, Durham, NC 27708, USA

Natasa S. Vidic
Department of Industrial Engineering, University of Pittsburgh, Pittsburgh, PA 15261, USA

Radisav D. Vidic
Department of Civil and Environmental Engineering,University of Pittsburgh, Pittsburgh, PA 15261, USA

Nathaniel R. Warner
Department of Earth Sciences, Dartmouth College, 6105 Fairchild Hall, Hanover, NH 03755, USA

Tieyuan Zhang
Department of Civil and Environmental Engineering, University of Pittsburgh, Pittsburgh, PA 15261, USA; and National Energy Technology Laboratory (NETL), Pittsburgh, PA 15236, USA

Introduction

During the past decade, advances in technologies critical to cost-effectively producing natural gas and oil from impermeable shale formations—hydraulic fracturing, seismic imaging, and horizontal drilling—have together caused a dramatic increase in both proved reserves and production of oil and gas in the United States. Energy development in U.S. shale plays began with experimentation and then a flood of new natural gas production in the Barnett Shale in Texas and Oklahoma. The Marcellus Shale, in the mid-Atlantic, was close behind; natural gas production from the Marcellus has risen from about 1 billion cubic feet per day in 2007 to about 16 billion cubic feet per day in early 2015 (U. S. Energy Information Administration 2015). U.S. oil production from shales and other "tight" formations more than tripled between 2010 and 2014, with the Bakken (North Dakota) and the Eagle Ford (Texas) shales leading the way (U.S. Energy Information Administration 2014). Though significant increases like these have been focused in the United States, production increases may follow in countries such as China and Argentina (U.S. Energy Information Administration 2013).

The increase in U.S. oil and gas production due to hydraulic fracturing (HF) and its complementary technologies has been a boon to many local and regional economies, and to U.S. consumers in the form of low energy prices (Mason et al. forthcoming). At the same time, researchers and policymakers have raised concerns about the impacts of this process on water resources, due to the water intensity of HF, its significant wastewater treatment and disposal burden, and the possibility of accidental surface releases and groundwater contamination (Vengosh et al. 2014).

A typical shale gas or tight oil well is drilled vertically, like conventional oil and gas wells, but is then supplemented by a set of wells drilled laterally from the bottom of the vertical wellbore through the targeted portion of the formation, significantly increasing the surface area of the formation in contact with the wellbore relative to conventional wells. The horizontal portions of the well are then hydraulically fractured; water, sand, and chemical additives (including scale inhibitors, friction reducers, and biocides to maintain well integrity) are injected into the well at very high pressure, focusing on one horizontal section at a time, fracturing the formation, propping open the created fractures, and allowing hydrocarbons to flow up the wellbore once the pressure is reduced.

Conventional oil and gas production often produces significant wastewater flows, since these resources often co-exist with ancient brines—often referred to as "produced water". Unconventional production compounds this problem by adding significant water inputs, some of which—estimates vary by shale play and range from 10 to 40 percent—also returns to the wellhead as "flowback" before production begins (Gregory et al. 2011, Jia et al. 2013). The quantity of water used in HF and the quantity that returns to the surface as flowback depend on geology, total well production, the number of fracture stages, and many other factors specific to well locations and the firms engaged in well completion. Operators in the Marcellus Shale use 2-4 million gallons for HF, on average, and those in the Barnett use about 5 million gallons (Veil 2010, Nicot et al. 2014). Both flowback and produced water typically contain high concentrations of dissolved salts and other solids, volatile organic compounds, naturally occurring radioactive material, and heavy metals, though many of these constituents have higher concentrations in produced water than in flowback (Vengosh et al. 2014).

There are many avenues through which HF could potentially impact water quality. Accidental releases of fracking fluids, flowback and produced water in transit or from inadequate storage and disposal facilities; groundwater contamination due to faulty well casing and cementing or via geological conduits between fractured wells and overlying aquifers; erosion and sedimentation from land clearing and well construction; and the release of partially treated flowback and produced water to rivers and streams have all been the subjects of both scientific study and discussions in the popular media since the beginning of the recent boom in unconventional oil and gas development. While all of these potential risks are worthy of additional research, the last of these concerns is the major focus of this book.

Wastewater flows from shale gas development in the Marcellus by 2012 represented more than a five-fold increase over baseline oil and gas wastewater flows in 2004, before unconventional production began in earnest (Lutz et al. 2013). Flowback and produced water in the Marcellus has had three main destinations—recycling for re-use in HF for additional wells, transport to deep injection wells primarily located in eastern Ohio, and treatment by municipal sewage treatment plants and industrial chemical waste treatment (CWT) facilities in Pennsylvania and, to a lesser extent, surrounding states. The relative importance of these re-use and disposal mechanisms has shifted over time. In the mid-2000s, little of this wastewater stream was recycled; now, most is recycled, though produced water continues to pose a more significant challenge than flowback for re-use (Jiang et al. 2014). Through 2011, treatment by municipal sewage treatment plants was common, but by 2012, the Commonwealth of Pennsylvania had eliminated this practice due to concerns regarding the ability of these facilities to remove many shale gas liquid waste constituents, particularly dissolved solids. The fraction of flowback and produced water sent to CWTs for treatment and release to rivers and streams has dropped, as well, though these facilities are still important disposal mechanisms in the Marcellus (Zhang et al. 2014).

The literature now contains significant evidence, from independent groups of researchers using very different analytical techniques and experimental locations, that the disposal to rivers and streams of partially-treated shale gas flowback and produced water has damaged water quality. Statewide in Pennsylvania, average chloride concentrations have been significantly elevated downstream of facilities treating shale gas wastewater (Olmstead et al. 2013). Bromide concentrations in rivers in southwest Pennsylvania near Pittsburgh have been elevated by this practice, even impacting the quality of finished drinking water drawn from the affected sources (Wilson and Van Briesen 2013). Radionuclides are accumulating in stream sediments directly downstream of CWTs treating flowback and produced water from the Marcellus (Warner et al. 2013a).

These demonstrated surface water quality impacts from the release of partially-treated shale gas wastewater to rivers and streams are serious, but they are regional in nature. Other major U.S. shale plays have plentiful deep injection well capacity—for example, at the end of 2014, Texas had more than 7,500 active disposal wells used for this purpose, while Pennsylvania had only seven. A paucity of geologically suitable locations for such wells in Pennsylvania, as well as public resistance to and a long timeline for permitting new disposal wells, suggest that operators in the Marcellus will continue to face a significant waste disposal challenge for the foreseeable future. As in Pennsylvania, the impact of HF on surface water quality globally may depend critically on the availability of effective treatment and disposal mechanisms for these high volume, highly concentrated liquid wastes.

The articles in Part I of this volume together illustrate the scope of the shale/wastewater challenge. Nicot et al. (2014) use self-reported data from energy operators to describe water withdrawals for HF, reuse, and flowback and produced water disposal patterns in the Barnett Shale. The paper quantifies the water intensity of HF, the typical source of withdrawals (groundwater vs. surface water), the fraction of flowback and produced water re-used, the time profile of flowback and produced water output, and the magnitude of flows to disposal wells – all important issues related to the impact of HF on water scarcity, and to the wastewater disposal challenge. Jia et al. (2013) analyze the fate of injected water, using simulation to understand how the fraction of water that remains in the formation after well completion may vary with geological and engineering factors. This subject is of critical interest to operators interested in recycling and optimizing well performance, as well as regulators and researchers working to understand the magnitude of potential risks to groundwater and the scale of the wastewater disposal challenge. Barbot et al. (2013) tackle an area in which data have been particularly sparse— the quality of flowback and produced water. Pooling together samples collected by the authors, those reported by operators to an industry association, and prior analyses by Pennsylvania's state environmental regulatory agency, the authors describe flowback and produced water constituents, and their changes over geographic space and time. If future wastewater flows in the Marcellus and other regions with limited deep injection well capacity are to be managed, additional research examining the quality of these flows is sorely needed.

Part II of the volume examines what is known about the environmental effects of flowback and produced water from unconventional wells, focusing on water resource impacts. Olmstead et al. (2013) perform the first large-scale analysis of HF impacts on rivers and streams, finding no systematic evidence of contamination from accidental releases of flow-back and produced water at well sites, but identifying two water quality concerns of importance—increases in downstream chloride concentra-tions from the treatment and disposal of shale gas wastewater by munici-pal sewage treatment plants and CWTs, and increases in total suspended solids (TSS) concentrations downstream from well pads, likely due to land clearing and pad construction. Warner et al. (2013b) explore a different po-tential water contamination pathway, comparing flowback water samples and private drinking water well samples in Arkansas' Fayetteville Shale to examine whether methane and brine from hydraulically fractured wells might contaminate local groundwater, but do not find evidence consistent with significant contamination. Brown (2014) catalogs the potential envi-ronmental impacts of naturally occurring radioactive material (NORM) present in deep shale formations that returns to the surface with flowback and produced water. This has been a particular concern in the Marcellus Shale, the most radiogenic of U.S. shale plays. Brown summarizes the potential risks from HF-related NORM, including occupational exposure from scaling of well equipment, storage of waste containing NORM in surface impoundments, accumulation in stream sediments from wastewa-ter treatment, risks to the integrity of deep injection wells, and disposal of NORM-containing solid waste at landfills.

Part III of the volume includes two articles that consider the potential for mitigation of risks associated with liquid wastes from unconventional wells. Using well characteristics and reservoir properties from a single Marcellus well in Pennsylvania, Olawoyin et al. (2012) model the use of novel technologies and design properties that recover 86% of injected fluids (much higher than the standard 10-40% for Marcellus wells), inte-grated with centralized flowback treatment by forward osmosis for re-use. Increasing fluid recovery rates and developing cost-effective and environ-mentally effective treatment processes is essential to the future manage-ment of the shale wastewater challenge. Zhang et al. (2014) examine the potential for one common treatment process used by CWTs—sulfate pre-cipitation—to remove the NORM typically present in Marcellus brines treated by these facilities. They conclude that the use of this typical treat-ment process will create a solid waste stream with much higher NORM concentrations than those allowable in municipal landfills, and is likely contributing to buildup of NORM concentrations in surface waters receiving treated Marcellus wastewater. Importantly, they suggest alternative treatment and disposal processes that would better address the radiogenic properties of Marcellus wastewater.

The three articles in Part IV begin to address the complicated issue of potential regulation and risk management. Hatzenbuhler and Centner (2012) survey the federal, regional, and state environmental statues and policies that establish oversight roles for regulators at various levels of government for the Marcellus region. At the end of their analysis, the authors describe potential additional approaches, including changes to federal law governing HF due to the National Energy Policy Act of 2005, and mandatory disclosure of HF fluid chemicals. Finally, the excerpt from Gosman (2013) takes up this last issue—fracking fluid disclosure. The U.S. Department of the Interior has proposed rules requiring such disclosure for operators working on federal land, though they have not been adopted. In the meantime, many states have adopted their own fracking fluid disclosure requirements. Gosman (2013) identifies 22 states with these requirements, and six in which they are under consideration. The author describes variation in state approaches to disclosure, classifying them into three distinct models, and considers both the "virtue and the perils" (p. 87) of disclosure as a response to the potential risks from HF fluids. The article by Chen and colleagues (2014) offers a summary of the issues discussed in the rest of the book, connecting the scientific, environmental, regulatory, and economic factors involved.

The tension between economic benefit and potential environmental consequences from using HF to extract oil and natural gas from shales and other tight formations has increased during the past 10 years of extraction in the United States, and is poised to become a global concern. A growing body of research suggests several current and potential negative environmental impacts of this process, and wastewater treatment and disposal has already proven to be one of the most significant challenges in the Marcellus Shale, where deep injection capacity is limited. The articles collected in this volume offer a survey of the evidence for the most important real and potential impacts, as well as some useful technical and policy solutions. The collection forms an important foundation for the ongoing academic research that should guide regulatory policy, and paired with effective communication of results, shape public opinion, as well.

REFERENCES

1. Barbot, Elise, Natasa S. Vidic, Kelvin B. Gregory, and Radisav D. Vidic. 2013. Spatial and Temporal Correlation of Water Quality Parameters of Produced Waters from Devonian-Age Shale following Hydraulic Fracturing. Environmental Science and Technology 47: 2562-2569.
2. Brown, Valerie J. 2014. Radionuclides in Fracking Wastewater: Managing a Toxic Blend. Environmental Health Perspectives 122(2): A50-A55.
3. Chen, Jiangang, Mohammed H. Al-Wadei, Rebekah C. M. Kennedy, and Paul D. Terry. 2014.Hydraulic Fracturing: Paving the Way for a Sustainable Future? Journal of Environmental and Public Health 2014.
4. Darrah, Thomas H., Avner Vengosh, Robert B. Jackson, Nathaniel R. Warner, and Robert J. Poreda. 2014. Noble Gases Identify the Mechanisms of Fugitive Gas Contamination in Drinking-Water Wells Overlying the Marcellus and Barnett Shales. Proceedings of the National Academy of Sciences 111(39): 14076-14081.
5. Gosman, Sara R. 2013. Reflecting Risk: Chemical Disclosure and Hydraulic Fracturing. Georgia Law Review 48: 83-144.
6. Gregory, K. B., R. D. Vidic, and D. A. Dzombak, 2011. Water Management Challenges Associated with the Production of Shale Gas by Hydraulic Fracturing. Elements 7(3): 181-186.
7. Hatzenbuhler, Heather, and Terence J. Centner. 2012. Regulation of Water Pollution from Hydraulic Fracturing in Horizontally-drilled Wells in the Marcellus Shale Region, USA. Water 4(4): 983-994.
8. Jackson, Robert B., Avner Vengosh, Thomas H. Darrah, Nathaniel R. Warner, Adrian Down, Robert J. Poreda, Stephen G. Osborn, Kaiguang Zhao, and Jonathan D. Karr. 2013. Increased Stray Gas Abundance in a Subset of Drinking Water Wells Near Marcellus Shale Gas Extraction. Proceedings of the National Academy of Sciences 110(28): 11250-11255.
9. Jia, Hongtao, John McLennan and Milind Deo. 2013. The Fate of Injected Water in Shale Formations. In: Effective and Sustainable Hydraulic Fracturing, ed. Andrew P. Bunger, John McLennan, and Rob Jeffrey, pp. 807-815, DOI: 10.5772/56443, available at: http://cdn.intechopen.com/pdfs-wm/44676.pdf.
10. Jiang, Mohan, Chris T. Hendrickson, and Jeanne M. Van Briesen. 2014. Life Cycle Water Consumption and Wastewater Generation Impacts of a Marcellus Shale Gas Well. Environmental Science and Technology 48(3): 1911-1920.
11. Lutz, B.D., A.N. Lewis, and M.W. Doyle. 2013. Generation, Transport, and Disposal of Wastewater Associated with Marcellus Shale Gas Development. Water Resources Research 49(2): 647-656.

12. Mason, Charles F., Lucija A. Muehlenbachs, and Sheila M. Olmstead. 2015. The Economics of Shale Gas Development. Annual Review of Resource Economics, in press, DOI: 10/1146/annurev-resource-100814-125023.

13. Nicot, Jean-Philippe, Bridget R. Scanlon, Robert C. Reedy, and Ruth A. Costley. 2014. Source and Fate of Hydraulic Fracturing Water in the Barnett Shale: A Historical Perspective. Environmental Science and Technology 48(4): 2464-2471.

14. Olawoyin, Richard, Christian Madu, and Khaled Enab. 2012. Optimal Well Design for Enhanced Stimulation Fluids Recovery and Flowback Treatment in the Marcellus Shale Gas Development using Integrated Technologies. Hydrology Current Research 3: 141.

15. Olmstead, Sheila M., Lucija A. Muehlenbachs, Jhih-Shyang Shih, Ziyan Chu, and Alan J. Krupnick. Shale Gas Development Impacts on Surface Water Quality in Pennsylvania. 2013. Proceedings of the National Academy of Sciences 110(13): 4962-4967.

16. Olmstead, Sheila M., and Nathan Richardson. 2014. Managing the Risks of Shale Gas Development Using Innovative Legal and Regulatory Approaches. William & Mary Environmental Law and Policy Review 39(1): 177-199.

17. Osborn, Stephen G., Avner Vengosh, and Nathaniel R. Warner. 2011. Methane Contamination of Drinking Water Accompanying Gas-Well Drilling and Hydraulic Fracturing. Proceedings of the National Academy of Sciences 108(20): 8172-8176.

18. Richardson, Nathan, Madeline Gottlieb, Alan Krupnick, and Hannah Wiseman. 2013. The State of State Shale Gas Regulation, RFF Report, June. Washington, DC: Resources for the Future. Available at: http://www.rff.org/rff/documents/RFF-Rpt-StateofStateRegs_Report.pdf, accessed 15 February 2015.

19. U.S. Energy Information Administration. 2013. Technically Recoverable Shale Oil and Shale Gas Resources: An Assessment of 137 Shale Formations in 41 Countries outside the United States, Washington, DC. Available at: www.eia.gov/analysis/studies/worldshalegas, accessed 14 February 2015.

20. U.S. Energy Information Administration. 2014. Annual Energy Outlook 2014. Washington, DC: U.S. EIA. Available at: www.eia.gov/forecasts/aeo/pdf/0383(2014).pdf, accessed 14 February 2015.

21. U.S. Energy Information Administration. 2015. Marcellus Region Drilling Productivity Report. Available at: www.eia.gov/petroleum/drilling/pdf/marcellus/pdf, accessed 14 February 2015.

22. Veil, J. Water Management Technologies Used by Marcellus Shale Gas Producers, Final Report, Argonne National Laboratory, U.S. Department of Energy, Argonne, IL.

23. Vengosh, Avner, Robert B. Jackson, Nathaniel Warner, Thomas H. Darrah, and Andrew Kondash. 2014. A Critical Review of the Risks to Water Resources from Unconventional Shale Gas Development and Hydraulic Fracturing in the United States. Environmental Science and Technology 48(15): 8334-8348.

24. Warner, Nathaniel R., Cidney A. Christie, Robert B. Jackson, and Avner Vengosh. 2013a. Impacts of Shale Gas Wastewater Disposal on Water Quality in Western Pennsylvania. Environmental Science and Technology 47(20): 11849-11857.

25. Warner, Nathaniel R., Timothy M. Kresse, Phillip D. Hays, Adrian Down, Jonathan D. Karr, Robert B. Jackson, and Avner Vengosh. 2013b. Geochemical and Isotopic

Variations in Shallow Groundwater in Areas of the Fayetteville Shale Development, North-Central Arkansas. Applied Geochemistry 35: 207-220.

26. Wilson, Jessica M., and Jeanne M. Van Briesen. 2013. Source Water Changes and Energy Extraction Activities in the Monongahela River, 2009-2012. Environmental Science and Technology 47(21): 12575-12582.

27. Zhang, T. et al. 2014. Co-precipitation of Radium with Barium and Strontium Sulfate and its Impact on the Fate of Radium during Treatment of Produced Water from Unconventional Gas Extraction. Environmental Science and Technology 48(8): 4596-4603.

Sheila Olmstead, PhD

PART I

WATER USE
AND WASTEWATER PRODUCTION
IN SHALE GAS DEVELOPMENT

CHAPTER 1

Source and Fate of Hydraulic Fracturing Water in the Barnett Shale: A Historical Perspective

JEAN-PHILIPPE NICOT, BRIDGET R. SCANLON, ROBERT C. REEDY, AND RUTH A. COSTLEY

1.1 INTRODUCTION

Hydraulic fracturing (HF) has become a hotly debated topic, particularly in regard to the volume of water used and the potential for aquifer contamination. (1, 2) Although HF and horizontal drilling has been practiced for decades, the combination of the two resulted in the exponential increase in gas production from <1% of U.S. gas production in the early 2000s to 40% in 2012 (3) (9.6 Tcf; 9.6×10^{12} standard cubic feet; 272 Gm^3). With expansion of HF into more water-scarce regions in the western U.S. and potential expansion into semiarid regions globally, understanding the volume of water required for HF is particularly important. Even in more humid settings, water availability can be an issue during droughts. Previous studies estimated HF water use for Texas [2011: 81.5 thousand AF (kAF), 100.2 million m^3 (Mm^3), including shales and tight formations,

Reprinted with permission from: Nicot J-P, Scanlon BR, Reedy RC, and Costley RA. Environmental Science and Technology *48,4 (2014). DOI: 10.1021/es404050r. Copyright 2014 American Chemical Society.*

SI section] and in Colorado (2011: 15 kAF/yr, 18.5 Mm³/yr). (4, 5) An estimated 13.2 kAF (16.3 Mm³) was used for HF in Oklahoma in 2011. (6) Although these water-use estimates represent a small fraction of water used in each state (~0.1% in Colorado, ~0.5% in Texas, and <0.5% in Oklahoma), the volumes may be significant locally, depending on competition with other sectors. Additional water-use estimates are available for the Marcellus Shale, totaling 32 kAF (39 Mm³) consumed between June 2008 and the end of 2012 in the Susquehanna River Basin, mostly in 2011–2012 (15–20 kAF/yr; 18–25 Mm³/yr) (7) and 23.5 kAF (29 Mm³) within the 2008–2012 period in the Upper Ohio River Basin, mostly in 2011–2012 (8.4 kAF/yr; 10.4 Mm³/yr). (8) Water demand in the Bakken area is estimated to be ~22 kAF/yr (~27 Mm³/yr). (9, 10)

Understanding the source of the water used for HF is important to assess the impact on water resources. To date, much of the water used has been fresh water from surface water or groundwater sources. Plays in more humid regions generally rely on surface water, whereas limited surface water availability in more semiarid regions may result in more groundwater use. The Marcellus Shale play uses predominantly surface water controlled by different river basins, e.g., the Susquehanna and Delaware basins. (11) In contrast, the Eagle Ford play lies mostly in a semiarid region and relies heavily on groundwater from the Carrizo-Wilcox aquifer because of limited surface water availability. (4)

The amount of HF water that flows back to the surface, commingled with water from the formation (produced water), termed flowback-produced (FP) water (see SI), is important because it controls the absolute volume that can be reused or recycled or the volume that must be disposed. (12, 13) Reuse is generally understood as requiring little treatment, whereas recycling suggests more involved treatment. (12) Shale formations (Marcellus (14) and Eagle Ford (15)) traditionally have been described as having small volumes of FP water.

Disposal approaches vary by play. Piping and trucking to centralized facilities for treatment and reuse is dominant in the Marcellus Shale with some on-site operations, (14, 16, 17) but injection wells (see SI) are preferred in the Barnett, (4) Eagle Ford, (15) and Bakken (10, 18) areas, despite improving technological capabilities in using high-salinity waters (50 000 mg/L and higher total dissolved solid (TDS)). (19)

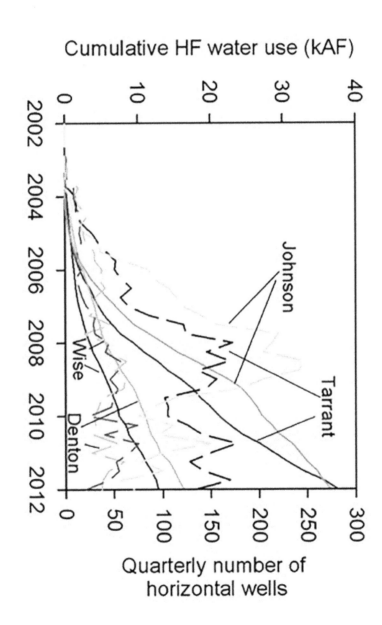

FIGURE 1: Cumulative water use for horizontal wells and their quarterly well count in the four counties of the core area (Denton, Johnson, Tarrant, and Wise; see SI).

The Barnett Shale play provides an ideal case for assessment of issues related to production of unconventional resources such as shale gas or shale oil. The Barnett Shale area (~26 000 mi², 68 000 km²) occupies ~45 counties in Central and North Texas, extending from suburban to rural settings (Figure S2). This study focuses on a ~10 000 mi² (~26 000 km²) area in ~15 counties in the eastern area of the shale footprint with hydrocarbon production potential. (20-23) It includes the core area (Figure 1) and most of the development activity. It was the first shale play in the world to be fully developed with HF (23, 24) and be subjected to intense HF. After a start in 1981 and through the 1990's with vertical wells, the combined use of horizontal drilling and HF that originated in the play in the early 2000s has allowed economical recovery of gas from shales. Operational details related to HF as applied to the Barnett Shale have been described. (19-23) The Barnett Shale produced an average of 1.9 Tcf/yr in the 2008–2012 5-year period (2.1 Tcf in 2011), to be compared to a total U.S. gas production of 28.5 Tcf in 2011, including 8.5 Tcf from shale gas wells. (3) Cumulative production since 1993 totals 14.9 Tcf as of April 2013. (25) Total production, including past and projected production, has been estimated at 45.1 Tcf. (26)

Gas production began in the mid-1990s using vertical wells and transitioned in 2003–2005 to mostly horizontal wells. Following a period of strong growth in the mid-2000s (>2000 wells/yr), drilling declined in the late 2000s because of reduced demand following an economic slump toward the end of the decade and decreasing natural gas prices. Although drilling activity has abated at the edges of the play core, it is still vigorous in the core itself (26, 27) and has increased in the so-called combo play (combined oil and gas production) in the northern portion of the play, in Cooke and Montague counties where HF-enhanced oil production has increased sharply since mid-2010.

The objective of this study was to assess the amount of water used for HF and the sources of that water, followed by an analysis of FP water and of its fate, including evaluation of disposal through injection and recycling (Figure S1). This study builds on previous work (4) that quantified HF water use in all Texas shale plays up to mid-2011 by increasing spatial resolution, increasing temporal resolution from annual to quarterly, extending analysis from water use to disposal and reuse, and assessing reliability of results by interviewing operators.

1.2 MATERIALS AND METHODS

1.2.1 WATER USE FOR HYDRAULIC FRACTURING

Data on water use were obtained from the commercial IHS database, (28) which, in turn, is based on water use that is self-reported by operators to the Railroad Commission (RRC), the state regulatory agency for oil and gas activities in Texas. Building on Nicot and Scanlon, (4) the analysis time period extends through December 2012. The analysis focuses on 2000 and the following years, as pre-2000 water use is <1 kAF (<1 Mm³). Data reporting from Barnett Shale operators is high, with >90% of wells reporting water volume, proppant amount, and lateral length of wells providing multiple checks on the reported water-use data. Water-use intensity (water volume used per unit length of lateral), proppant loading (proppant mass per unit water volume), and mean and median values were used to detect reporting errors. (4, 29) Similar information is available from the Web site FracFocus (http://fracfocus.org), but, as of August 2013, not in a format that can be readily queried and, more importantly, FracFocus only includes data from 2010, precluding retrospective analysis.

1.2.2 SOURCE OF WATER FOR HYDRAULIC FRACTURING

The source of water for HF is more difficult to access than amount of water used, because no regulation requires reporting of water sources. Therefore, we relied on a mix of hard data and soft data such as interviews to provide estimates. The industry generally uses water sources that are most readily available and economic for a given time and location. Sources can be classified into unequivocal (1) surface water and (2) groundwater, with several other minor categories of either ultimate origin and in decreasing importance: (1) municipal water from either urban reservoirs or water hydrants; (2) recycling/reuse of HF water, of treated industrial or municipal wastewater; and (3) small, distributed sources such as farm ponds. The information can be obtained from either the users (industry) or the water suppliers. We interviewed several major operators in the play about their practices relative to water sourcing in 2012 (30) and again in 2013.

Water suppliers include self-suppliers, local landowners, municipalities, larger water districts, and river authorities with various levels of reporting and data accessibility. The first two groups (self-suppliers and landowners) rely mostly on groundwater, whereas the last two groups use surface water mostly. Information on groundwater use is generally obtained from groundwater conservation districts (GCDs, see SI). The study area contains five multicounty GCDs (Figure S2) out of ~100 in the state, all but one created within 2007–2009 (Table S1); therefore, only very recent data are potentially available. Whereas groundwater is owned by the landowner and withdrawals are controlled by the rule of capture with some restrictions posed by GCDs, surface water use follows a prior appropriation doctrine ("first in time, first in right") and is owned and strictly regulated by the state, which grants permits and regulates the resource. As such, volumes of surface water withdrawn are well-known but their ultimate use is not, because several uses are bundled into larger categories, e.g., in the case of HF, "mining." River authorities are state entities that manage their respective river basins and operate reservoirs and treatment plants. They also hold some water rights. Four river authorities (Figure S2) could potentially provide water to the oil and gas industry. We contacted GCDs, river authorities, water districts, and several municipalities (Fort Worth, Arlington) in the course of this study.

1.2.3 HYDRAULIC FRACTURING WATER QUALITY

Overall, public-domain information on ionic composition of HF water is qualitative at best. Water quality is not reported to the RRC. Some companies report TDS, but not the ionic makeup of HF fluids, to FracFocus. Operator interviews provided additional information. Water quality can be inferred from some sources, e.g., surface water and wastewater treated to state standards being fresh.

1.2.4 FLOWBACK-PRODUCED WATER CHARACTERISTICS

RRC regulations require that operators report oil, gas, and water production on a monthly basis. Although operators perform routine chemical analyses on an as-needed basis, TDS and ionic makeup of FP water are not recorded systematically and very few data sets are available in the public domain. Production water volumes were compiled from the IHS database. (28) About 10% of the wells do not have production water data, most likely because of lack of reporting, and is consistent with the fraction of wells with no reported HF water use. We examined a total of 12 228 horizontal wells.

1.2.5 INJECTION OF FP WATER FOR DISPOSAL

Information about injection volumes is accessible through both the IHS database (28) and the RRC Web site. The RRC has the apparent benefit of singling out disposal from HF operations, whereas IHS provides information about individual wells but not the source of the injected water. The RRC regulates U.S. EPA Class II wells and has for many years been tracking water injected for disposal and water used for waterflooding and reservoir pressure maintenance. Injection can be done by commercial entities, which manage wells disposing of oil and gas waste and salt water into nonproducing intervals, or by oil and gas companies, which operate the vast majority of Class II wells.

In Texas, most FP water is disposed of into injection wells—information that has recently (end of 2011) become specifically available from the RRC. (31) In the past, reporting of Class II injection from HF operations was combined with conventional (not HF) salt-water disposal. The Texas Class II injection well count is ~50 000; ~20% of these are disposal wells—i.e., injecting into nonproducing formations. A query of the IHS database for Class II wells in the 15-county area yielded ~2000 wells. Fluid injection into 1383 wells was reported during the period from 2001 through 2012. Unlike production, which must be reported on a monthly basis, injection volumes are reported to the RRC only annually; therefore, injection volumes in this study are accurate only if reported before and during summer 2012.

1.3 RESULTS AND DISCUSSION

Overall, hydrocarbon production is fragmented among ~250+ operators but dominated by a few companies. According to the IHS database, a total of 17 685 horizontal and vertical wells reported in the play at the beginning of 2013 were operated by 250+ companies (see SI).

1.3.1 WATER USE

1.3.1.1 HISTORICAL WATER USE AND CONSUMPTION

Barnett Shale water use in 2011 totaled ~25.8 kAF, amounting to ~32% of the total HF water use in Texas in 2011, including HF in tight formations, (30) and down from a high of 28.8 kAF in 2008 (Figure S3). Until the end of 2002, wells were mostly vertical and restricted predominantly to Denton and Wise counties (with a cumulative total of ~3.8 and ~3.6 kAF), out of a cumulative total of 8.3 kAF. The estimated total amount of water used in the play to the end of 2012 is ~170 kAF, including ~152 kAF for horizontal wells (Figure S4a,b) and an additional ~18 kAF for vertical wells. Tarrant and Johnson counties are the largest water users (Figure 1). Water use increased outward from the core area until 2008, contracted back to the core area in 2009, and then shifted toward the combo play to the north and the liquid-rich area to the northwest (Figures S5 and S6).

Water use is currently almost exclusively related to HF of horizontal wells, which peaked in 2008 with fracturing of ~2750 horizontal wells. The peak year for HF of vertical wells occurred in 2002 with a total of ~750 wells. Horizontal wells account for the bulk of the water use and the length of the laterals has been slowly increasing in the past few years (median ~3800 ft in 2011), with a concomitant increase in water use per well (Figure 2a, Figure S4c; Figure S7). Water use is often reported on a per-well basis, and, in the case of the Barnett Shale, water use per well has increased from ~3 Mgal/well in mid 2000s to ~5 Mgal/well in 2011 (1 Mgal = 3.8 thousand m^3). However, increasing trends in water use per well are misleading because they reflect an almost doubling of the lengths of laterals during that time. A more useful indicator is normalized water use per length of lateral or water-use intensity, which has remained steady at ~1100–1200 gal/ft (1.4–1.5 m^3/m) since 2007 (Figure 2b). Note that, in the years 2003–2006, water-use intensity was generally much higher but was steadily decreasing, finally stabilizing when operators perfected the HF technology in horizontal wells; a total of ~2300 horizontal wells were completed to the end of 2006 vs an estimated 10 500 wells from that point to the end of 2012. In contrast to water-use intensity, proppant loading has been increasing over time, from 0.2 lb/gal in 2002 to ~0.8 lb/gal in 2009 (25 to 100 kg/m^3), plateauing until the beginning of 2012, and slightly decreasing since then (Figure S4d).

Water consumption is different from water use. In this work, water use is defined as the amount of water required to perform HF stimulations, whatever the source of the water. Water consumption is defined as the amount of fresh water abstracted from surface water or groundwater. Most water used in the Barnett Shale is estimated to be consumed; operator interviews reveal that ~23.7 kAF (~92% of total water use) was consumed in 2011. Additional HF water (~5%) is derived from reuse/recycling of used-water streams. The remainder (~3%) consists of brackish water originating from mostly brackish water sections of aquifers.

The Barnett Shale water use represents a small fraction (0.14%) of total statewide water use (18.1 million AF in 2011) as reported by the Texas Water Development Board (TWDB). Water use in Texas is reported in terms of withdrawal for all categories and consumption for thermoelectric generation. Total water use has averaged 15.4 million AF/yr for 2005 through 2011, with interannual variations related mostly to irrigation needs. Statewide water consumption has been estimated at 10.2 million AF in 2010 (32) and 11.4 million AF on average for the period 2005–2011 (13.4 million AF in 2011 translating into 0.18% for Barnett Shale water consumption). When analyzed at the county level, HF water use can represent a much higher fraction, especially in rural, sparsely populated counties (Figure 3). However, water may originate from outside the county, particularly in large population centers (see below). Water for auxiliary uses, e.g., for drilling, is relatively small and strongly operator-dependent. For example, some operators use oil-based muds requiring very little water, while others use water-based muds potentially requiring up to 0.5 Mgal/well but more often <0.25 Mgal/well. (29, 33)

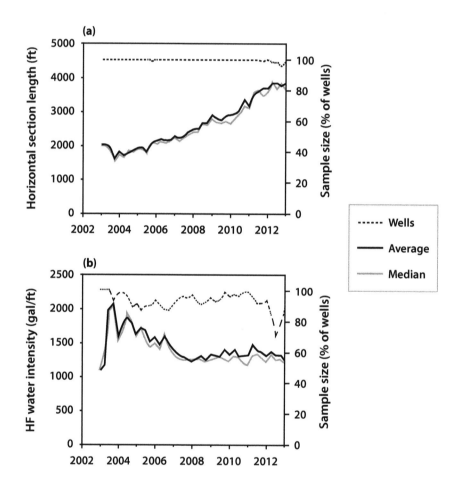

FIGURE 2: Data on Barnett Shale horizontal wells, including various historical parameters and coefficients for reported and estimated water use as a function of time: (a) average/median lateral length and fraction of wells for which it is reported; (b) average/median water-use intensity and fraction of wells for which both HF water use and lateral length are reported. Tick marks on the x-axis represent the beginning of the year. Other parameters are reported on Figure S4 (number of wells completed per quarter and cumulative count; cumulative water use; average/median water use per well; and average/median proppant loading).

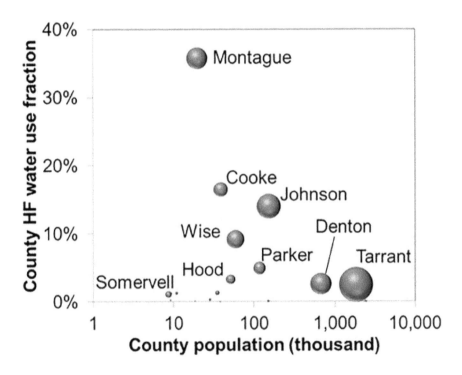

FIGURE 3: 2010 County population vs 2011 fraction of water use in the county for HF purposes. Bubble size is related to absolute HF water use (for example, 8.8 kAF in Tarrant County, 1.5 kAF in Cooke County, and 0.3 kAF in Somervell County).

Water-use intensity is higher in Denton County and in the eastern half of Wise County, where the Barnett Shale is deepest and also where many older horizontal wells are located (Figure S8). High water-use intensity in Montague County most likely reflects early production from the oil window. The cumulative length of laterals in a given area or county (Figure S9) can be used to estimate the average well density (Figure S10). Density of well laterals is fairly high in Johnson County and the southern half of Tarrant County. The county with the highest relative cumulative length of laterals (Johnson County) yields an average spacing between assumed parallel laterals of ~1700 ft (Table S2). This value is much greater than the operational distance between laterals of ~1000 ft or even 500 ft, (33, 34) suggesting that Johnson County, despite its past HF activity, is still likely to see further significant drilling and HF activity, as illustrated by the coverage gaps (Figure S8). The decrease in well completion activity in Johnson County (Figure 1) is more related to gas prices than to true depletion of the resource.

1.3.1.2 SOURCE AND QUALITY OF WATER FOR HYDRAULIC FRACTURING

Data on the source of HF water are sparse. The industry is fragmented, and within the same company, practices may differ from one lease to the next and through time. Water contracts are signed and expire in a very dynamic business environment, suggesting that collected information can only be considered semiquantitative. Available data suggest that the play as a whole relies roughly equally on both groundwater and surface water. At least three temporal phases are discernible, with the middle phase relying more on surface water but all relying strongly on fresh water. During the initial phase, up to 2006, groundwater was estimated at 50%+ of total water consumption. (35) Interviews suggest that, during the second phase, 2007–2010, operators used more surface water, estimated at 70–80% of water consumed during that period, (30) but with considerable variations among operators and locations. A plausible explanation for such a pattern resides in the typical approach followed by operators. Water-supply wells initially tap local groundwater unless the stimulated well is close to surface water. Then, after the initial period during which operators drill to hold leases (often 3 years) and explore for sweet spots (areas of high gas production), exploration and production become more predictable, and semipermanent water lines are installed from surface water reservoirs that can provide large amounts of water at relatively low cost. The third phase (from 2011) shows a renewed reliance on groundwater related to development of the combo play in Montague and Cooke counties. Montague County groundwater use increased from ~1 kAF in 2009 to 5.4 kAF in 2011.

Groundwater is derived mostly from the Trinity aquifer, (36) the only major aquifer underlying 76% of the 15-county area. A large fraction of Trinity aquifer withdrawals are for municipal use. (36, 37) The aquifer is one of the most depleted aquifers in the state. (38, 39) The underlying Paleozoic aquifer system (15) supplies some water in Montague County. GCDs provided volumes for some or all HF-related groundwater withdrawals for years 2011–2012 (Figure S11 and Table S3). They account for more than half (~15 kAF/yr) of the annual total HF water use.

Most of the 15-county area of interest is located in the Brazos (51%) and Trinity (46%) river basins. The combined reservoir conservation-pool storage capacity in the 15-county area is 2700 kAF. The Trinity River Authority does not supply water to oil and gas operators. The Brazos River Authority, with the largest watershed, has contracts with operators to provide HF water but data on water deliveries are only available for the broader mining category, which includes HF water use. The Brazos River Authority delivered an increasing water volume from 2001 (2.6 kAF) to 2008 (5.7 kAF), but has supplied 2.1 kAF/yr (in 2012) or less since then in the mining category, following the general trend of HF water use. On the basis of this pattern, it is logical to assume that most of the mining-category water use is for HF.

HF water can also originate far from the Barnett Shale footprint. As is often the case in large urban centers, water is imported from distant reservoirs to provide water to municipal and industrial customers. (40) Such is the case in Tarrant County (which includes the City of Fort Worth), with the primary water supplier being the Tarrant Regional Water District (TRWD) (41) providing water to many municipalities in the county and operating large reservoirs southeast of Dallas (Figure S12). A significant fraction of the Tarrant County HF water use is provided directly by TRWD and was as high as 3.5 kAF in 2009, decreasing to 1.0 kAF in 2012. The remaining surface water sources include smaller water providers and unknown surface water right holders. Municipalities (Arlington, Fort Worth, Dallas) also provide water directly to operators, either through direct withdrawals from urban reservoirs before water is treated, or as treated water through hydrants (>4 kAF in 2011). In both cases, the ultimate water source is from the municipal supply. Tarrant County has the highest water use in the play, both annual and cumulative (Figure 1); however, HF water use is nevertheless a very small fraction of total water use (Figure 3).

Interviews with operators hinted that some use water from brackish aquifers, (30) estimated to be ~3% of HF water use and highest in the combo play in Montague County and on the western edges of the play. The Trinity aquifer (42) and the north-central Texas Paleozoic aquifers (15) contain slightly brackish horizons interspersed with fresher horizons. However, the largest source of salinity comes from blending fresh water with FP water. Some operators also use outflow from wastewater treatment plants. Texas Commission on Environmental Quality (TCEQ) has records, but no volumes, showing that some treated wastewater from large cities (Dallas, Fort Worth, Waco) and smaller towns (Bowie, Cisco, Keene, Weatherford) is used.

Overall, less recycling/reuse and brackish water use is currently occurring in the Barnett than in other Texas plays further west or south. (29, 30) A large operator in 2005–2011 processed 2.24 kAF of FP water to generate 1.6 kAF of pure water to be used for stimulating new wells. Knowing that this particular operator manages ~21% of the wells and has had a more active recycling program than most operators, we extrapolated that the entire field used ~7.7 kAF of recycled water (1.6 kAF/21%); i.e., 5.5% of the total of 139 kAF of water used in that period. Interviews with operators are consistent with this estimate, suggesting that ~5% of HF water is from reuse/recycling for the past few years. Periodic droughts, characteristic of Texas climate, do not seem to control HF water use in the Barnett play, which is more sensitive to the price of gas and economic activity (Figure S13).

1.3.2 FP—FLOWBACK/PRODUCED WATER

FP water flow decreases rapidly with time after wells are allowed to produce. Records from the IHS database show that percentiles (5th to 90th) of monthly water production steadily decline over time (Figure S14). Percentiles also show large variability, with a median for maximum monthly production <5000 bbl/month (0.64 AF/month; 1 bbl = 0.159 m^3) but a 90th percentile >20 000 bbl/month (2.58 AF/month) and a 5th percentile of ~0 bbl/month. However, cumulative production can still result in large volumes: median ~75 000 bbl (9.67 AF) after 4 years with a 90th percentile >300,000 bbl (38.7 AF) but a 5th percentile of 7000 bbl (0.90 AF) (Figure S15). A more interesting metric is the ratio of FP water to the amount used for HF, which we call the HF water balance ratio (WB ratio) (Figure 4). After one year, the median WB ratio is ~60%, but the mean is >100% because of a few wells with exceptional water production. After several years, the median exceeds 100% of HF water. The variability of the ratio is large (Figure 4), ranging from 20% (5th percentile) to 350% (90th percentile) after 4 years. In interviews, operators tended to underestimate the amount of FP water as reported by IHS, likely focusing on the initial period during which some treatment and recycling can still take place. At later times, monthly volumes are small, but cumulatively they amount to a non-negligible fraction of the overall FP water.

The spatial distribution of the county-level medians of the WB ratios is not random but structured, with a minimum in the core area increasing outward (Figures S16 and S17), suggesting that the higher the amount of HF water retained in the shale, the higher the gas production. It is, however, premature to draw a direct causal relationship. Efforts are underway to relate this observation to various gas-production parameters, including the so-called maturing or soaking time, during which a well remains shut-in after HF, and geological parameters, e.g., porosity, pore-size distribution, and rock competence. A water-encroachment operational explanation, in which the underlying karstic Ellenburger Formation is systematically breached during HF, is unlikely; the Viola-Simpson Formation pinch-out (20, 21) does not seem to control the WB ratio. A time-dependence of the WB ratio, suggesting possible operational improvements through time, is not clear: percentiles in Tarrant and Denton counties trend in opposite directions over time (Figure S18). The WB ratio does not appear to be related to operator skill level: comparing WB ratios from different large operators where leases are commingled shows no significant difference. Note that producing less water in the core area means that less water is available for reuse/recycling. Quality and chemical composition of the FP water are only known through anecdotal evidence. (43, 44)

1.3.3 INJECTION WELLS

Injection-well count (all vertical) has increased in the Barnett Shale play during the past decade. Until 2002, HF was confined to Denton and Tarrant counties and all injection activities outside of these counties were related to conventional hydrocarbon production (Figure S19, year 2000). Injection activity in Cooke and Montague counties, the NW half of Wise County, Jack and Palo Pinto counties, and the western half of Parker County is clearly related to conventional oil production. All wells active in 2000 in this area with no change or decrease in injection volumes are assumed to be unrelated to FP water and other HF spent-fluid disposal. The Ellenburger Formation that underlies the Barnett Shale is the injection horizon of choice, (45) although FP water was also reinjected into shallower formations above the Barnett Shale in the early years.

Within the 15-county area, 8.8 kAF/yr of water was injected in 2000, representing the baseline in the NW corner of the area. In 2011, the injection rate had increased 5-fold in ~10 years to 45.7 kAF/yr; i.e., ~36.9 kAF/yr attributed to HF activities through ~150 currently active commercial injection wells. (46) A significant fraction of disposal occurs in Johnson County, which has the highest injection-well count (Figure S20) and receives more than twice the volume of water to be disposed than the county listed second (Parker County) (Figures 5, S21 and S22). A cumulative total of 170 kAF has been disposed of through injection wells from 2000 through 2011, whereas a total of 152 kAF was used in HF operations (Figure S23), although the latter number can be reduced by 5%, to 144 kAF, to account for recycled/reused water. (30) This result is consistent with the observation that many Barnett wells produce back >100% of the volume injected (Figure 4) and with the understanding that many wells have been

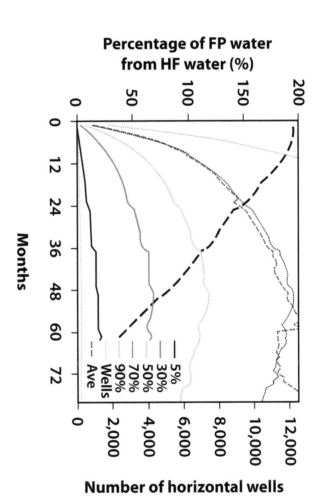

FIGURE 4: WB ratio; that is, ratio of FP water to HF water over time (5th, 30th, 50th, 70th, 90th percentiles, and average) and number of wells having data (dotted line). The base for the calculation includes only horizontal wells. The monthly records of each well were sequentially ordered from the first month where water was produced to the last month of record, specifically ignoring initial months with zero water production. For all wells for a given month, percentiles were then calculated. Logically, the number of wells with many months of production is much lower than the number of wells with a few months of production, because many wells were completed recently.

fractured only recently and will produce significant amounts of water unless shut in. Natural evaporation from storage pits could also reduce the
volume of fluids to be injected. (13)

Injection of FP water in the Barnett Shale area represents <4% of the
wastewater volume injected in Texas each year. Statewide injection volume for a 12-month period (Oct. 2011 to Sept. 2012) was 924 kAF, similar to the previously reported value of 951 kAF for 2007. (47) Note that a
small fraction of the injection wells are thought to have produced seismic
events strong enough to be felt at the surface, (48, 49) but the HF operation itself has not been documented as generating felt seismic events. As
mentioned earlier, the RRC has recently started to report water disposal in
a specific HF category. However, the statewide volume of 6.0 kAF for the
same 12-month period used above shows that the current RRC data clearly
underestimate the volume of HF fluids disposed in injection wells, most
likely as a result of underreporting in the HF category and reporting to the
salt-water general category instead.

1.3.4 IMPLICATIONS AND THE FUTURE

Drilling activity in the Barnett Shale play has been decreasing since a peak
in 2008, despite periodic surges related to increased demand for natural
gas following renewed economic activity or interest in condensates and
oil. However, use of water for HF has remained relatively steady since
2009 because the mean lateral length has almost doubled. The price of gas,
which steadily increased between 2002 and 2008 to ~$8/McfHH (thou-
sand cubic feet Henry Hub; 1 Mcf = 28.3 m^3) and higher (Figure S13d)
and then quickly dropped to $2–$4/McfHH, translated into a focus in the
core area which is likely to continue. The current average well spacing and
projection of drilling activity (26) suggest sustained drilling for several
decades. Broadly, groundwater and surface water each account for half
of the new HF water with periodic swings favoring one or the other. HF
water needs in the core area will be met by local groundwater resources,
in particular, the confined section of the Trinity aquifer, but also, very
likely, by imported water. As the Dallas/Fort Worth area grows, it secures
large contracts for water that originates from outside the metroplex. It then

makes sense to project that operators in the core area will keep acquiring water from local surface water districts and will be able to meet HF water needs, especially when combined with recycling/reuse and use of brackish water. Currently, most of the FP water is disposed through deep well injection. Given that injected water volumes are larger, on average, than HF volumes, growth in recycling/reuse is possible.

A metric that has been proposed to assess competing water uses is known as the water intensity or the amount of water (gal or m^3) used to produce a unit volume of gas (Mcf or m^3) or unit of energy (MBtu or GJ). Because all water use is up front during well completion, (4) the ultimate water intensity depends on total gas production and the estimated ultimate recovery. With most wells still producing, this number is not yet accessible, although more and more data are in the public domain (Figure S24). The current water intensity can be computed from cumulative HF water use and gas production (Figure S25); at the end of 2012, it reached 4.37 gal/Mcf (15.7 L/GJ), which is clearly an upper bound. Extrapolating trends from older wells yields more accurate values (Figure S26) and suggests that, after 6 years of production, the water intensity is in the range of 2.5–3 gal/Mcf (9.0–10.8 L/GJ), consistent with findings by Clark et al. (50) This value would then evolve downward over time, to the range of 1–2 gal/Mcf (3.6–7.2 L/GJ).

REFERENCES

1. Jackson, R. E.; Gorody, A. W.; Mayer, B.; Roy, J. W.; Ryan, M. C.; Van Stempvoort, D. R.Groundwater Protection and Unconventional Gas Extraction: The Critical Need for Field-Based Hydrogeological Research Groundwater 2013, 51 (4) 488–510

2. Vidic, R. D.; Brantley, S. L.; Vandenbossche, J. M.; Yoxtheimer, D.; Abad, J. D., Impact of Shale Gas Development on Regional Water Quality. Science 2013, 340 (6134).

3. EIA. Natural Gas Summary. http://www.eia.gov/dnav/ng/ng_sum_lsum_dcu_nus_a. htm (accessed August 2013).

4. Nicot, J.-P.; Scanlon, B. R.Water Use for Shale-Gas Production in Texas, U.S Environ. Sci. Technol. 2012, 46 (6) 3580–3586

5. COGAWater Sources and Demand for the Hydraulic Fracturing of Oil and Gas Wells in Colorado from 2010 through 2015, 2012; p 9; http://water.state.co.us/DWRIPub/ CGWC%20Meetings%20and%20Process%20Documents/Oil%20and%20Gas%20 Water%20Sources%20Fact%20Sheet%20-%20Final.pdf.

6. Murray, K. E.State-Scale Perspective on Water Use and Production Associated with Oil and Gas Operations, Oklahoma, U.S Environ. Sci. Technol. 2013, 47 (9) 4918–4925

7. Richenderfer, J. Water Acquisition for Unconventional Natural Gas Development Within the Susquehanna River Basin (July 1, 2008 thru Dec. 31, 2012). http://www2.epa.gov/sites/production/files/documents/richenderfer.pdf (accessed August 2013).

8. Mitchell, A. L.; Small, M.; Casman, E. A.Surface Water Withdrawals for Marcellus Shale Gas Development: Performance of Alternative Regulatory Approaches in the Upper Ohio River Basin Environ. Sci. Technol. 2013, 47 (22) 12669–12678

9. Schaver, R. Water Management Issues Associated with Bakken Oil Shale Development in Western North Dakota. http://www.mgwa.org/meetings/2012_fall/plenary/shaver.pdf (accessed August 2013).

10. NDDMR (North Dakota Industrial Commission Department ofMineral Resources) Oil & Gas Activity Update, June 20, (2013) https://www.dmr.nd.gov/oilgas/presentations/presentations.asp (accessed August 2013).

11. Arthur, J. D.; Uretsky, M.; Wilson, P. In Water Resources and Use for Hydraulic Fracturing in the Marcellus Shale Region; http://www.netl.doe.gov/technologies/oil-gas/publications/ENVreports/FE0000797_WaterResourceIssues.pdf, 2010 Marcellus Shale: Energy Development and Enhancement by Hydraulic Fracturing Conference, Pittsburgh, PA, May 5–6, AIPG - American Institute of Professional Geologists: Pittsburgh, PA, May 5–6, 2010; p 22.

12. Horner, P.; Halldorson, B.; Slutz, J. A. Shale Gas Water Treatment Value Chain - A Review of Technologies, including Case Studies; SPE-147264; p 10; SPE Annual Technical Conference and Exhibition; Society of Petroleum Engineers; Denver, Colorado, USA, 2011.

13. Slutz, J. A.; Anderson, J. A.; Broderick, R.; Horner, P. H. Key Shale Gas Water Management Strategies: An Economic Assessment Tool; SPE-157532, p 15; In International Conference on Health, Safety and Environment in Oil and Gas Exploration and Production, 2012; SPE/APPEA International Conference on Health, Safety, and Environment in Oil and Gas Exploration and Production; Perth, Australia; 2012.

14. Wilson, J. M.; VanBriesen, J. M.Oil and Gas Produced Water Management and Surface Drinking Water Sources in Pennsylvania Environmental Practice 2012, 14 (04) 288–300

15. Nicot, J.-P.; Huang, Y.; Wolaver, B. D.; Costley, R. A.Flow and Salinity Patterns in the Low-Transmissivity Upper Paleozoic Aquifers of North-Central Texas GCAGS J. 2013, 2, 53–67

16. Lutz, B. D.; Lewis, A. N.; Doyle, M. W.Generation, Transport, And Disposal of Wastewater Associated with Marcellus Shale Gas Development Water Resour. Res. 2013, 49 (2) 647–656

17. Maloney, K. O.; Yoxtheimer, D. A.Production and Disposal of Waste Materials from Gas and Oil Extraction from the Marcellus Shale Play in Pennsylvania Environmental Practice 2012, 14 (04) 278–287

18. Nicot, J.-P.; Duncan, I. J.Common Attributes of Hydraulically Fractured Oil and Gas Production and CO_2 Geological Sequestration Greenhouse Gases-Science and Technology 2012, 2 (5) 352–368

19. King, G. E., Hydraulic fracturing 101: What every representative, environmentalist, regulator, reporter, investor, university researcher, neighbor and engineer should know about estimating frac risk and improving frac performance in unconventional gas and oil wells; SPE-152596, pp 651– 730. In SPE Hydraulic Fracturing Technology Conference 2012; Society of Petroleum Engineers; The Woodlands, TX, United States; February 6–8, 2012.

20. Montgomery, S. L.; Jarvie, D. M.; Bowker, K. A.; Pollastro, R. M.Mississippian Barnett Shale, Fort Worth basin, North-Central Texas: Gas-Shale Play with Multi-Trillion Cubic Foot Potential AAPG Bull. 2005, 89 (2) 155– 175

21. Pollastro, R. M.; Jarvie, D. M.; Hill, R. J.; Adams, C. W.Geologic Framework of the Mississippian Barnett Shale, Barnett-Paleozoic Total Petroleum System, Bend Arch-Fort Worth Basin, Texas AAPG Bull. 2007, 91 (4) 405– 436

22. Zhao, H.; Givens, N. B.; Curtis, B.Thermal Maturity of the Barnett Shale Determined from Well-Log Analysis AAPG Bull. 2007, 91 (4) 535– 549

23. Bruner, K. R.; Smosna, R. A Comparative Study of the Mississippian Barnett Shale, Fort Worth Basin, and Devonian Marcellus Shale, Appalachian Basin; U.S. DOE NETL report No DOE/NETL-2011/1478; URS Corporation; http://www.netl.doe.gov/technologies/oil-gas/publications/brochures/DOE-NETL-2011-1478%20Marcellus-Barnett.pdf, 2011; p 106.

24. Martineau, D. F.History of the Newark East Field and the Barnett Shale as a gas reservoir AAPG Bull. 2007, 91 (4) 399– 403

25. RRC. Newark, East (Barnett Shale) Total Natural Gas 1993 through May 2013; http://www.rrc.state.tx.us/barnettshale/NewarkEastField_1993-2013.pdf (accessed August 2013).

26. Browning, J.; Ikonnikova, S.; Gülen, G.; Tinker, S.Barnett Shale Production Outlook, SPE-165585 SPE Economics & Management 2013, 5 (3) 89– 104

27. Browning, J.; Tinker, S. W.; Ikonnikova, S.; Gülen, G.; Potter, E.; Fu, Q.; Horvath, S.; Patzek, T.; Male, F.; Fisher, W.; Roberts, F.; Medlock, K., IIIBarnett Shale Model – 1: Study develops decline analysis, geologic parameters for reserves, production forecast Oil & Gas Journal 2013, 111 (8) 63– 71

28. IHS. Enerdeq - Energy Information Access and IntegrationPlatform; http://www.ihs.com/products/oil-gas-information/data-access/enerdeq/index.aspx (accessed August 2013).

29. Nicot, J.-P.; Hebel, A.; Ritter, S.; Walden, S.; Baier, R.; Galusky, P.; Beach, J. A.; Kyle, R.; Symank, L.; Breton, C. Current and Projected Water Use in the Texas Mining and Oil and Gas Industry: The University of Texas at Austin, Bureau of Economic Geology 2011, 357http://www.twdb.state.tx.us/publications/reports/contracted_reports/doc/0904830939_MiningWaterUse.pdf

30. Nicot, J.-P.; Reedy, R. C.; Costley, R.; Huang, Y. Oil & Gas Water Use in Texas: Update to the 2011 Mining Water Use Report 2012, 97http://www.twdb.state.tx.us/publications/reports/contracted_reports/doc/0904830939_2012Update_MiningWaterUse.pdf

31. RRC online system, H10 Filing System; http://webapps.rrc.state.tx.us/H10/h10PublicMain.do (accessed August 2013).

32. Scanlon, B. R.; Reedy, R. C.; Duncan, I. J.; Mullican, W. F.; Young, M. H.Controls on Water Use for Thermoelectric Generation: Case Study Texas, U.S Environ. Sci. Technol. 2013, 47 (19) 11326– 11334

33. Gong, X.; McVay, D.; Bickel, J. E.; Montiel, L. V., Integrated Reservoir and Decision Modeling To Optimize Northern Barnett Shale Development Strategies; SPE-149459; p 13; Canadian Unconventional Resources Conference, Society of Petroleum Engineers; Alberta, Canada, 2011.

34. Mutalik, P. N.; Gibson, R. W., Case History of Sequential and Simultaneous Fracturing of the Barnett Shale in Parker County; SPE-116124; p 7; SPE Annual Technical Conference and Exhibition; Society of Petroleum Engineers; Denver, Colorado, USA, 2008.

35. Nicot, J.-P.; Potter, E. Barnett Shale Groundwater Use Estimates, Appendix 2; In Northern Trinity/Woodbine aquifer groundwater availability model: assessment of groundwater use in the northern Trinity aquifer due to urban growth and Barnett Shale development; Bené, P. G., Harden, R., Griffin, S. W., and Nicot, J.-P. ;The University of Texas at Austin, Bureau of Economic Geology, 2007; p 81; http://www.twdb.state.tx.us/groundwater/models/gam/trnt_n/TRNT_N_Barnett_Shale_Report.pdf.

36. Bené, P. G.; Harden, R.; Griffin, S. W.; Nicot, J.-P. Northern Trinity/Woodbine aquifer groundwater availability model: assessment of groundwater use in the northern Trinity aquifer due to urban growth and Barnett Shale development; Harden & Associates, Austin, TX, contract report to the Texas Water Development Board, 2007; http://www.twdb.state.tx.us/groundwater/models/gam/trnt_n/TRNT_N_Barnett_Shale_Report.pdf.

37. Burk, R. A.; Kallberg, J.Rule of Capture and Urban Sprawl: A Potential Federal Financial Risk in Groundwater-Dependent Areas International Journal of Water Resources Development 2012, 28 (4) 659– 673

38. TWDB. 2007 State Water Plan; http://www.twdb.state.tx.us/waterplanning/swp/2007/index.asp; 2007.

39. Mace, R. E.; Dutton, A. R.; Nance, H. S. In Water-Level Declines in the Woodbine, Paluxy, and Trinity Aquifers of North-Central Texas, Gulf Coast Association of Geological Societies (GCAGS) Transactions, XLIV, 1994; pp 413– 420.

40. Fry, M.; Hoeinghaus, D. J.; Ponette-Gonzalez, A. G.; Thompson, R.; La Point, T. W.Fracking vs Faucets: Balancing Energy Needs and Water Sustainability at Urban Frontiers Environ. Sci. Technol. 2012, 46 (14) 7444– 7445

41. TRWD. Tarrant Regional Water District, Overview; http://www.trwd.com/AboutUs (accessed August 2013).

42. Chaudhuri, S.; Ale, S.Characterization of Groundwater Resources in the Trinity and Woodbine Aquifers in Texas Sci. Total Environ. 2013, 452–453, 333– 348,

43. Acharya, H. R.; Henderson, C.; Matis, H.; Kommepalli, H.; Moore, B.; Wan, H. Cost Effective Recovery of Low-TDS Frac Flowback Water for Re-use; U.S. DOE NETL report No DE-FE0000784; GE Global Research, Niskayuna, NY; 2011; http://www.netl.doe.gov/technologies/oil-gas/publications/ENVreports/FE0000784_FinalReport.pdf.

44. Hayes, T. D.; Severin, B. F. Characterization of Flowback Waters from the Marcellus and the Barnett; RPSEA Report No. 08122–05.09, Gas Technology In-

stitute, Des Plaines, IL; 2012; http://barnettshalewater.org/documents/08122-05.09CharacterizationofFlowbackWaters3-16-2012.pdf.

45. Ficker, E., Five Years of Deep Fluid Disposal into the Ellenburger of the Fort Worth Basin, Search and Discovery; Article #80227; AAPG 2012 Southwest Section Meeting; Ft. Worth, Texas, 2012.

46. RRC. Disposal/Injection Well Counts by District, Field, Lease; http://www.rrc.state.tx.us/data/wells/waterbank.php (accessed August 2013).

47. Clark, C. E.; Veil, J. A. Produced water volumes and management practices in the United States; Argonne National Laboratory report ANL/EVS/R-09/1, 2009; http://www.veilenvironmental.com/publications/pw/ANL_EVS__R09_produced_water_volume_report_2437.pdf.

48. Frohlich, C.Two-year survey comparing earthquake activity and injection-well locations in the Barnett Shale, Texas Proc. Natl. Acad. Sci. U.S.A. 2012, 109 (35) 13934– 13938,

49. Frohlich, C.; Hayward, C.; Stump, B.; Potter, E.The Dallas-Fort Worth Earthquake Sequence: October 2008 through May 2009 Bulletin of the Seismological Society of America 2011, 101 (1) 327– 340

50. Clark, C. E.; Horner, R. M.; Harto, C.Life cycle water consumption for shale gas and conventional natural gas Environ. Sci. Technol. 2013, 47 (20) 11829– 11836

There are several supplemental files and one figure that are not available in this version of the article. To view this additional information, please use the citation on the first page of this chapter.

CHAPTER 2

The Fate of Injected Water in Shale Formations

HONGTAO JIA, JOHN MCLENNAN, AND MILIND DEO

2.1 INTRODUCTION

The growth in producing hydrocarbons from unconventional reservoirs (shales) has been phenomenal. The production of liquids from the Eagle Ford play grew to about 52 million barrels in 2011 [1] (Figure 1).

The growth in production is driven by improvements in hydraulic fracturing technology. Multistage fracturing using long horizontal wells is the common practice. Millions of gallons of water are pumped into the formation to create these fractures. Industry data reveals that only about a third of the injected water is typically recovered. The fate of injected water is of fundamental interest. Use of large quantities of water in fracturing has brought into question the sustainability of this type of completion and development practice. Furthermore, low water recovery has prompted environmental concerns about whether the injected water leaves the target formation with a potential of infiltrating and contaminating aquifers. The purpose of this paper was to examine the capability of the formation to imbibe the injected water based on different capillary pressure relationships.

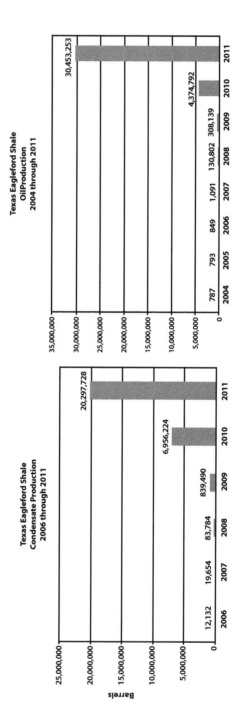

FIGURE 1: The phenomenal growth in production of liquids from shales with Eagle Ford. In just over a three-year period, insignificant production has been transformed to over 52 million barrels of liquids in 2011.

2.2 TECHNICAL APPROACH

The Advanced Reactive Transport Simulator (ARTS) at the University of Utah was used to perform simulation studies (Figure 2). ARTS is a modular reservoir simulator that has been under development over a number of years [2-4]. The main idea of ARTS is to decouple the discretization methods from the physical models. The discretization methods in ARTS include the conventional finite difference, control-volume finite element and a generalized control volume method. These discretization methods could be coupled with a variety of physical models. The simplest physical model would be simulation of a single-phase gas with immovable water phase. Two-phase and three-phase black oil models are used to simulate primary production followed by water and polymer flooding. Thermal processes such as steam flooding, in-situ combustion, steam-assisted gravity drainage, etc. are represented in K-value based thermal-compositional models. In these models, the vapor-liquid equilibrium is calculated using the ratio between the vapor and the liquid phase composition of each component (K-value). ARTS also includes a geochemical module to simulate processes associated with carbon dioxide sequestration and reactions involving carbon dioxide, brine and rocks.

The use of a control volume finite element model as one of the discretization schemes allows multiphase simulation of complex reservoir geometries including a discrete fracture network representation of natural and hydraulic fractures.

We represented and simulated two different discrete fracture domains in this work—both with non-orthogonal features (Figure 3). It is common practice to represent and simulate hydraulic fractures as orthogonal features. However, it is evident that the fractures created are not perfectly perpendicular to the horizontal well. The microseimic cloud that is observed in a number of cases with multiple horizontal fractures (for example, [5]), shows fractures that are more complex than regularly spaced orthogonal features. It is true that there is no one to one correlation between the microseismic signatures and the shape and morphology of hydraulic fractures. However, there are a number of indications that point to the hydraulic fractures being more complex than simple orthogonal features.

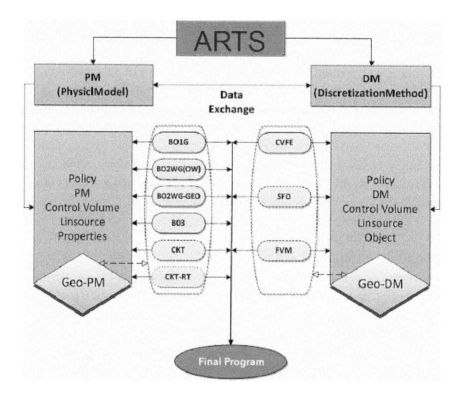

FIGURE 2: The framework used in simulating water injection and production in fractured systems. The discretization methods (DM) are decoupled from the physical models (PM).

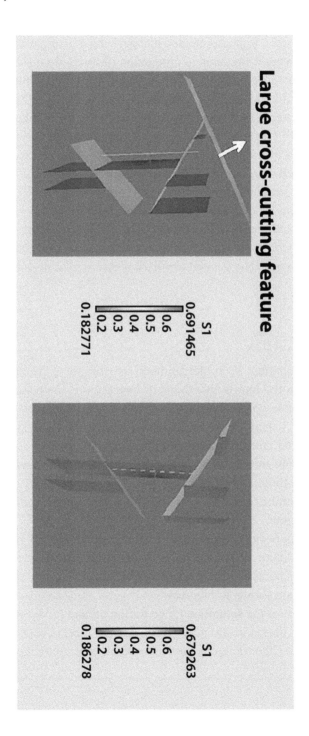

FIGURE 3: Figure showing two fractured systems simulated in this study.

The hydraulic fractures created interact with existing natural fractures. The role of natural fractures in production of fluids from shales is still an open question. The production behavior of both the gas and liquid reservoirs does not indicate a highly fractured system. On the other hand, when fracturing water is injected in a well, it is common to see interference in an adjacent well. This may be in the form of pressure interference or explicit breakthrough of water injected in the adjacent well. Pressure interference in and of itself does not indicate fluid transport to the well.

Capillary pressures for these shale reservoirs are not well characterized. The wettability of the reservoir rocks is also not well known. Al-Bazali et al. [6], measured sealing capacities of shale caprocks. This data provides some guidance for the capillary pressure values and relationships to use for these systems. The general capillary pressure relationship is given by:

$$P_c = 2\sigma\cos\theta r$$

In this equation, P_c is the capillary pressure, σ is the interfacial tension between the immiscible fluids of interest, θ is the contact angle and r is the average pore radius. Al-Bazali et al.[6], were considering shales that were less than 10 nD in permeability. For the three shales studied, they measured entry pressures ranging from 470 psia to 750 psia. They calculated pore throat radii of about 30 nM for entry pressures of crude oil. For pore throats of less than 10 nM (Sondergeld et al. [7]), very large capillary pressures (two to three times those measured by Al-Bazali et al [7]) are possible.

There has been much discussion about wettability of shales. In this paper, we examined the differences in water recovery due to variations in wettability of the rock. The three sets of oil-water capillary pressures used in the study are shown in Figure 4.

Over most of the saturation range for the oil and mixed wet situations, the capillary pressures are negative, indicating a preference for oil as the wetting fluid. Other domain-specific parameters are shown in Table 1.

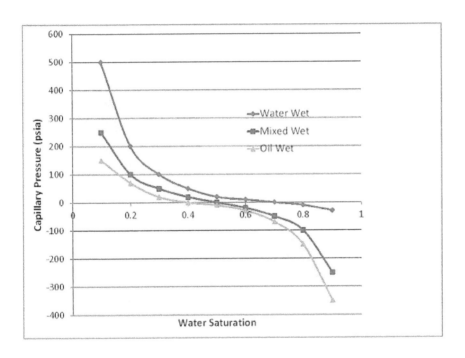

FIGURE 4: The three sets of capillary pressures used in this study.

TABLE 1: Properties of the domain and simulations

Domain Size	260 feet X 260 feet X 100 feet
Initial Reservoir Pressure	2000 psia
Fracture Permeability	1000 mD
Porosity	20%
Matrix Permeability	0.5 mD
Water Injected	30000 barrels

Water recovery after one month (30 days) for each of the simulations was compiled. For the base case capillary pressures, the water recoveries for the three wetting scenarios and for the two domains (one with the cross-cutting fracture, and one without) are shown in Table 2.

TABLE 2: Water recoveries for the three wetting scenarios and for the two domains studied in this paper. Recoveries are for the base case where the initial reservoir pressure was 2000 psia and the matrix permeability was 0.5 mD.

	Water Wet	Mixed Wet	Oil Wet
Water recovery ratio (With cross-cutting fracture)	21.53%	29.35%	36.28%
Water recovery ratio (Without the cross-cutting fracture)	22.97%	31.24%	38.39%

The water recoveries observed in the table above are consistent with water recoveries of about 20-40% listed in field observations. Water recoveries increase as we go from water wet to mixed wet to oil wet clearly indicating the tendency of the matrix to imbibe and hold water as the formation becomes more water wet. There is a 15% increase in water recovery as we go from water wet to the oil wet case. The presence of the long cross-cutting feature does not make a significant impact in recovery. The recovery does decrease as injected water is transported to longer distances—but the difference in recovery is only 1-2%.

In a number of shale reservoirs, the permeabilities are lower and the initial pressures are higher. To investigate the effects of these parameters on recovery, simulations were performed with 5000 psia initial pressure and 0.1 mD matrix permeability. Results of these simulations are tabulated in Table 3.

TABLE 3: Water recoveries for the three wetting scenarios and for the two domains studied in this paper. Recoveries are for the base case where the initial reservoir pressure was 5000 psia and the matrix permeability was 0.1 mD.

	Water Wet	Mixed Wet	Oil Wet
Water recovery ratio(With cross-cutting fracture)	37.42%	40.17%	44.19%
Water recovery ratio (Without the cross-cutting fracture)	41.02%	44.61%	49.83%

Higher initial pressure results in higher water recoveries, particularly in the water wet cases. The differences between recoveries with and without the large cross-cutting feature are now between 4-5%. The differences between the different wettability cases however are reduced to only about 8% (compared to about 15%) as the largest difference the water wet and the oil wet scenarios.

At smaller pore radii, the capillary pressures are expected to be larger. One set of simulations were performed where the shape of the base case capillary pressures were maintained, but the capillary pressures were increased ten times for each of the saturation values. The resulting recoveries are tabulated in Table 4.

TABLE 4: Water recoveries for the three wetting scenarios and for the two domains studied in this paper. Recoveries are for the case where the capillary pressures were ten times the base case capillary pressures used. The shapes of the capillary pressure curves were the same as the ones used in Figure 4. The initial reservoir pressure was 5000 psia and the matrix permeability was 0.1 mD.

	Water Wet	Mixed Wet	Oil Wet
Water recovery ratio(With cross-cutting fracture)	20.1%	27.15%	41.9%
Water recovery ratio (Without the cross-cutting fracture)	23.3%	30.2%	45.5%

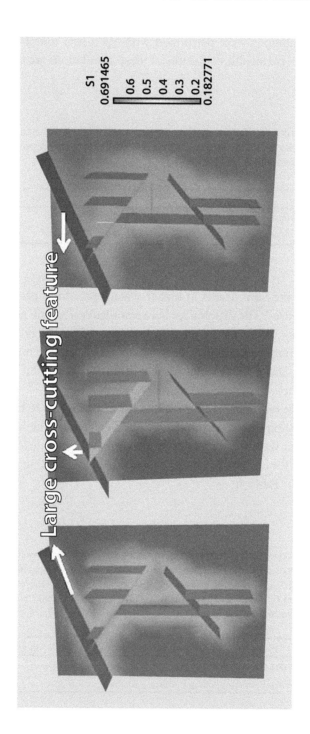

FIGURE 5: Figure showing water saturations in the matrix through one hydraulic fracture and interacting natural fractures. Left panel is for the water wet case, the middle panel is for the mixed wet case and the right panel is the oil wet case. As the wettability goes from water wet to oil wet the infiltration decreases increasing injected water recovery. In this particular example, the large cross-cutting feature does not take a significant amount of water off site.

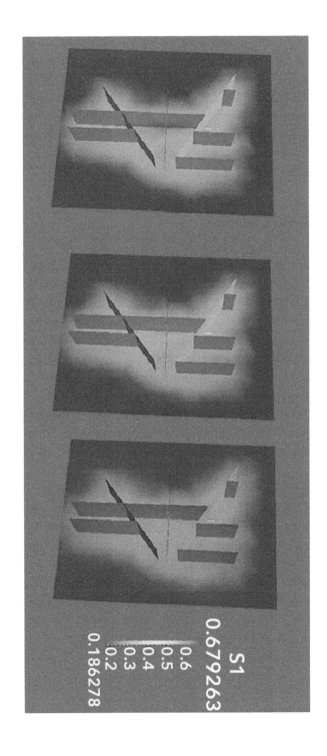

FIGURE 6: Figure showing water retained in the matrix through one hydraulic fracture and interacting natural fractures. Domain without the large cross-cutting feature is used. Left panel is for the water wet case, the middle panel is for the mixed wet case and the right panel is the oil wet case. Water saturation scale is also shown. As the wettability goes from water wet to oil wet the infiltration decreases increasing injected water recovery.

As the capillary pressure increases, more water is retained. For mixed wet and oil wet scenarios, water saturation in the matrix area is lower (Figure 5). Similar relative difference between recoveries is maintained when recoveries are compared for domains with and without the large cross-cutting features. The system without the large cross-cutting fracture in this case returns on the average about 3% more water than when the large fracture exists. Water saturations for the domain without the large fracture are shown in Figure 6.

2.3 CONCLUSIONS

Recovery of water injected for hydraulic fracturing in shales is only about 30%. There is a question of the fate of injected water. In this paper we studied water retention in shales for different shale wettability conditions. Two different domains where a hydraulic fracture intersected with a small existing network of natural fractures were used in the simulations. A specially developed framework that can handle representation of complex fracture networks was used for simulations. Capillary pressures in rocks containing very small pores tend to be high—of the order of 1000 psia. Three sets of capillary pressures—water wet, mixed wet and oil wet were examined. Simulations showed that a recovery of 20-30% is expected for typical water wet conditions, while a recovery of about 37%-48% is expected for oil wet scenarios. The recovery for mixed wet conditions fell between these two extremes. The recovery is reduced when a large cross-cutting fracture is introduced—but not significantly. That is because water will be recovered if the fractures are interconnected. Results discussed in this paper helped quantify the role of wettability in the recovery of water used for hydraulic fracturing. In this paper we assumed that the initial water saturation was low and that the water was immovable. If that is not the case, water saturation in the matrix and in the natural fractures, as well as the water-oil or water-gas relative permeability functions play significant roles in determining the water balance.

REFERENCES

1. Data from the Texas Railroad Commission- http://wwwrrc.state.tx.us/
2. Y. K Yang, 2003Finite-Element Multiphase Flow Simulator, Ph.D. dissertation, University of Utah, 2003.
3. Y Fu, 2007Multiphase Control Volume Finite Element Simulation of Fractured Reservoirs, Ph.D. dissertation, University of Utah, 2007.
4. Z Gu, 2010A Geochemical Compositional Simulator for Modeling C O2 Sequestration in Geological Formations, Ph.D. dissertation, University of Utah, 2010.
5. N. A Stegent, K Ferguson, and J Spencer, 2011Comparison of Frac Valves vs. Plug and Perf Completions in the Oil Segment of the Eagle Ford Shale: A Case Study, CSUG/SPE 148642, Paper presented at the Canadian Unconventional Resources Conference, Calgary, Canada, 1517November 2011.
6. T. M Al-bazali, J Zhang, M. E Chenevert, and M. M Sharma, 2005Measurement of the Sealing Capacity of Shale Caprocks, SPE 96100, Paper presented at the Annual Technical Conference and Exhibition, Dallas, Texas, Oct 912
7. M. E Curtis, R. J Ambrose, C. H Sondergeld, and C. S Rai, Structural Characterization of Gas Shales on the Micro- and Nano-scales, CSUG/SPE 137693, Paper presented at the Canadian Unconventional Resources and International Petroleum Conference held in Calgary, 1921October 2010

CHAPTER 3

Spatial and Temporal Correlation of Water Quality Parameters of Produced Waters from Devonian-Age Shale following Hydraulic Fracturing

ELISE BARBOT, NATASA S. VIDIC, KELVIN B. GREGORY, AND RADISAV D. VIDIC

3.1 INTRODUCTION

At a time when countries are trying to limit their carbon emissions in an effort to combat global warming trends, natural gas has emerged as a transitional fuel between coal-based energy and next generation, renewable sources. Domestic natural gas used in conventional power generation offers a potential environmental advantage over coal for power generation in terms of global warming potential.(1-3) The Marcellus Shale in the northern Appalachian basin (Figure 1) is an unconventional shale gas reservoir with vast development potential but also a source of concern for the potential environmental challenges that its exploitation may create.(4) In 2002, the undiscovered resources were estimated to be only 53.8 billion

Reprinted with permission from: Barbot E, Vidic NS, Gregory KB, and Vidic RD. Environmental Science and Technology *47*,6 (2013). DOI: 10.1021/es304638h. Copyright 2013 American Chemical Society.

m^3.(5) Engelder estimated that there is 50% probability that the Marcellus Shale will ultimately yield 13.8 trillion m^3 of natural gas.(6) This increase is mainly due to the emergence of horizontal drilling and hydraulic fracturing for well stimulation. In tight formations such as the Marcellus Shale, the density and dimensions (length and aperture) of natural fractures alone is not sufficient to achieve economical gas production. However, a combination of horizontal drilling and hydraulic fracturing greatly improves well productivity and economics of natural gas recovery from tight shales. Fracturing fluid is injected into the horizontal wellbore under high pressure (480–680 bar) to open and prop new and existing fractures in the formation. Hydraulic fracturing of a horizontal well creates a contact area that is thousands of times greater than that of a typical vertical well.(7) Fracturing fluid chemistry is designed according to the geological characteristics of each site and the chemical characteristics of the water supply used. The mix usually contains proppants, friction reducers, scale inhibitors, biocides, gelling agents and gel breakers, and an inorganic acid.(8, 9)

Upon the completion of hydraulic fracturing, the fluid is allowed to flow back to the surface to relieve the downhole pressure and allow gas migration to the surface. The term "flowback water" typically refers to the fracturing fluid mixed with formation brine flowing at high flow rate immediately following hydraulic fracturing and before the well is placed into production. "Produced water" then refers to the fluid that continues to be coproduced with the gas once the well is placed into production and may be present over the lifetime of the well. Chemical and physical characteristics of produced water from conventional and unconventional oil and gas reservoirs worldwide (including shale gas, conventional natural gas, conventional oil, coal-bed methane, and tight gas sands) and the potential treatment options for these waters have been extensively reported.(10-15) Recent study on Marcellus Shale flowback/produced water defines geochemical parameters that characterize the brine with the goal of tracing the water in case of leakage into other water bodies.(16)

Currently, most of the flowback water from Marcellus Shale in Pennsylvania is stored in surface impoundments or tanks and is then treated (e.g., filtration and/or precipitation for metal removal) to enable its reuse as a fracturing fluid. Flowback water reuse for hydraulic fracturing is an option chosen by an increasing number of oil and gas companies as it reduces water needs and wastewater management costs. High concentrations of calcium, barium, and strontium are an issue that can limit reuse because of the high scaling potential of these ions if the water were to be reused for hydraulic fracturing.(17) The objective of this study is to report the inorganic chemistry of flowback/produced water from the Marcellus Shale in the Appalachian basin and correlate it with spatial and temporal information. Data analyzed in this report are from water samples collected in this study as well as those that are available in the literature. Major cations, including barium, calcium, and strontium as the constituents of concern for water reuse, were analyzed and the flowback/produced water quality from Marcellus Shale was compared with other brines from adjacent formations. The evolution of salinity with time and location across Pennsylvania is also analyzed to provide additional insight into the origin and nature of salinity as it has a major impact on potential management strategies for this wastewater.

3.2 MATERIALS AND METHODS

3.2.1 FLOWBACK WATER SAMPLING

Flowback water samples have been collected at three well sites in southwest Pennsylvania (Sites A, B1, and B2). Site A includes 5 horizontal wells on a single pad that were hydraulically fractured within a short period of time using similar fracturing fluid. The fracturing fluid was a mix of flowback water from previously completed wells and fresh water, but the exact composition of the fluid was unknown. The five wells were fractured stage by stage simultaneously and the water flowed back to the surface at the same time from all five wells. Flowback water samples were collected at various times from day 1 to day 20 (day 1 referring to the first day the water was allowed to flow). On this particular site, the wells were shut in for 11 days between the end of the hydraulic fracturing event and the beginning of the flowback. Sites B1 and B2 are separated by 0.9 km and are characterized by a single well on a pad and no lag time between the end of the hydraulic fracturing and flowback. Hence, they present similarities in geographic location and in depth, length, number of stages and volume of fracturing fluid injected. Samples were collected from day 1 to 29 on site B1 and from day 1 to 16 on site B2.

3.2.2 ANALYTICAL METHODS

Total dissolved solids (TDS) and total suspended solids (TSS) were determined using the Standard Methods 2540C and 2540D, respectively. Alkalinity measurements were performed following the Standard Method 2320B.(18) Prior to cation analysis by atomic absorption (GBC908, GBC Scientific Equipment LLC, Hampshire, IL and Perkin-Elmer model 1000 AAS) the samples were filtered through a 0.45-µm nylon filter, acidified to pH below 2 using nitric acid, and kept at 4 °C. Samples for total iron analysis were prepared by dissolving the sample in 1 N H_2SO_4 before filtration. For Ca, Ba, and Sr analysis, the samples were diluted in 2% metal grade nitric acid and 0.15% KCl was added to the solution to limit ionization interferences. An air–acetylene flame was used for Na, Mg, and Fe analysis, while a nitrous oxide–acetylene flame was used for Ba, Sr, and Ca analysis to limit chemical interferences. Cation analysis was also performed by ICP-OES for several samples to verify AA methods. The two analytical methods were in very good agreement. Anions were analyzed using an ion chromatograph (Dionex DX-500) with a Dionex IonPac AS14A column at a flow rate of 1 mL/min.

3.2.3 OTHER DATA SOURCES

Several flowback/produced water data sources were used in this study, including the PADEP Bureau of Oil and Gas Management analyses of flowback/produced water from more than 40 sources(19) and the Marcellus Shale Coalition sampling and analyses of flowback water from several wells.(20) In addition, data shared by production companies were included in the analyses. For all samples, charge balance was checked and any sample exhibiting a charge balance below 85% was discarded. The location of sampled sites is indicated in Figure 1 that shows the number of wells sampled and the number of samples collected for each county.

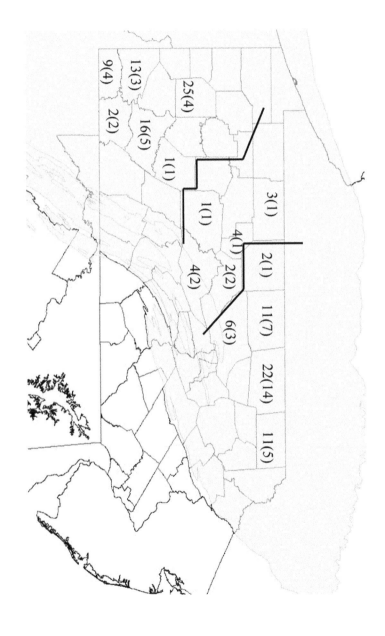

FIGURE 1: Map of Pennsylvania counties and underlying Marcellus Shale, with number of samples collected and in brackets the number of wells sampled [geospatial data from the USGS, available at www.pasda.psu.edu]. Black bold lines separate the Northeast, Central, and Southwest areas of the state.

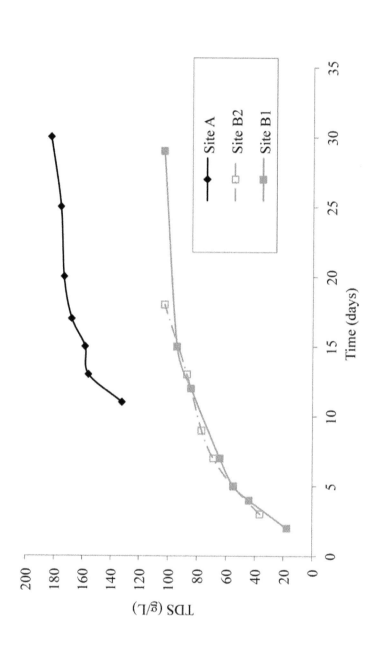

FIGURE 2: Variation of total dissolved solids concentration in flowback water versus time for sites A, B1, and B2. Day 0 corresponds to the end of the hydraulic fracturing process.

TABLE 1: Summary of Marcellus Shale Produced Water Quality in Pennsylvania

	minimum	maximum	average	number of samples
TDS (mg/L)	680	345,000	106,390	129
TSS (mg/L)	4	7,600	352	156
oil and grease (mg/L)	4.6	802	74	62
COD (mg/L)	195	36,600	15,358	89
TOC (mg/L)	1.2	1530	160	55
pH	5.1	8.42	6.56	156
alkalinity (mg/L as $CaCO_3$)	7.5	577	165	144
SO_4 (mg/L)	0	763	71	113
Cl (mg/L)	64.2	196,000	57,447	154
Br (mg/L)	0.2	1,990	511	95
Na (mg/L)	69.2	117,000	24,123	157
Ca (mg/L)	37.8	41,000	7,220	159
Mg (mg/L)	17.3	2,550	632	157
Ba (mg/L)	0.24	13,800	2,224	159
Sr (mg/L)	0.59	8,460	1,695	151
Fe dissolved (mg/L)	0.1	222	40.8	134
Fe total (mg/L)	2.6	321	76	141
gross alpha[a] (pCi/L)	37.7	9,551	1,509	32
gross beta[a] (pCi/L)	75.2	597,600	43,415	32
Ra^{228} (pCi/L)	0	1,360	120	46
Ra^{226} (pCi/L)	2.75	9,280	623	46
U^{235} (pCi/L)	0	20	1	14
U^{238} (pCi/L)	0	497	42	14

[a]*Data for Northeast Pennsylvania only.*

3.3 FLOWBACK/PRODUCED WATER CHARACTERIZATION

3.3.1 COMPOSITION OF FLOWBACK WATER RECOVERED WITH TIME

Flowback water is dominated by Cl–Na–Ca with elevated bromide, magnesium, barium, and strontium content and very low sulfate and carbonate.

Its chemistry varies greatly during the first weeks of collection. A summary of the key water quality parameters for samples examined in this study is presented in Table 1 and TDS profiles for the well sites sampled for this study are shown in Figure 2. Sites B1 and B2 exhibited much lower TDS content than site A, which is likely due to shorter residence time in the formation and the use of municipal drinking water as fracturing fluid rather than a mixture of freshwater and produced water as was the case for site A.

Chloride and sodium are the primary constituent ions, followed by calcium, barium, magnesium, and strontium, and their increase with time was similar to that of TDS (Figure 3). Concentration of strontium and magnesium in the flowback water from site A ranged from 1300 mg/L on day 11 to 2100 mg/L on day 30, while the concentration of barium reached only 380 mg/L on day 30. In contrast, barium concentration in the flowback water from sites B1 and B2 increased to 3000 mg/L on day 30, while magnesium concentration reached only 670 mg/L on the same day. For the two sites close to each other (site B1 and B2), the flowback water had similar concentration ranges for ions, demonstrating a strong correlation between geographic location and flowback water composition.

The decrease in pH and alkalinity with time (Figure 4a–b) as well as the decline of Ca/Mg ratio (Figure 4b) suggests precipitation of calcium carbonate within the formation. Equilibrium calculations were performed using the software Phreeqc and the Pitzer activity coefficient model. Calcite saturation index for Site A decreased from 0.83 to −0.15, indicating that this flowback water is essentially equilibrated with respect to calcium carbonate within 30 days. However, calcite saturation index for Site B1 ranged from 1.94 on day 2 to 1.02 on day 29, indicating that calcite continues to precipitate in the brine. Site A flowback water had much lower alkalinity in comparison with site B1. This difference may be due to a greater extent of calcium carbonate precipitation driven by the higher initial calcium content in Site A flowback water. Sulfate concentrations in samples collected from Site A were close to detection limit while sulfate concentrations in flowback water from site B1 decreased from 28.6 mg/L on the first day to 2 mg/L after 30 days. Reduction in sulfate concentration can be explained by barium sulfate precipitation and the fact that very little sulfate is present in the formation. Equilibrium calculations revealed that barite saturation index decreased from 2.15 on day 2 to 1.61 on day 29, confirming slow precipitation of $BaSO_4$. Such behavior is in agreement with previous studies that revealed fairly slow barite precipitation when the saturation index is below 2.6.(21)

3.3.2 ORIGIN OF SALINITY IN THE PRODUCED WATER

When injected in the wellbore, the fracturing fluid may mix with formation brine that exists in the target formation (Marcellus Shale in this case) or in adjacent formations should fractures extend outside the target formation. The Marcellus Shale is widely regarded as a dry formation, but there is a single report in the literature with three formation brine analyses.(22) The second salinity source can be the formation itself. XRD analysis of core samples revealed that shale from the Marcellus Shale is composed (by decreasing fraction) of quartz, clays, pyrite, and calcite.(23, 24) Blauch et al. describe salt layers containing calcium, sodium, potassium, iron, magnesium, barium, and strontium that may dissolve and contribute to salinity in flowback and produced water.(25) However, there are no other reports that describe salt layers, suggesting that they may not be present throughout the formation.

Inorganic constituents of produced water from the Marcellus Shale were compared with the Marcellus Shale formation brines described previously,(22) Bradford Formation brines located in the Upper Devonian,(22) and produced water from oil and gas wells in Western Pennsylvania from horizons ranging from Lower Silurian to Upper Devonian.(26) Most of the reports on produced water analyses from the Marcellus Shale used in this study are missing some critical information that is required for detail understanding of the produced water chemistry. For example, the PADEP Bureau of Oil and Gas Management data(19) provide exact sampling location but give no information about the time of contact between the water and the formation, or about the initial fracturing fluid quality. In addition, only one sample was collected per site, representing either the composition at a given time or mixed flowback/produced water collected over several days. The Marcellus Shale Coalition report(20) includes initial water quality and variation of flowback water composition with time but no information on the contact time with the formation. Despite the lack of precise information in these reports, the data can be used to analyze general trends in the geochemistry of produced water and provide information that is critical when evaluating potential management strategies for these wastewaters, especially in the Appalachian basin where water reuse for hydraulic fracturing is the preferred management alternative.

Chloride concentration was chosen as reference for other key ions as it is the major anion in flowback water and is strongly correlated with TDS independently of the location and sampling time ($R^2 = 0.90169$). The concentrations of key ions of interest (i.e., Na, Ba, Mg, Sr, Br) were compared to chloride concentrations for Marcellus Shale but the data were divided into 3 geographical zones: Southwest (SW), Central (C), and Northeast (NE) to assess the impact of geographic location on these correlations (Figure 1). Due to the small sample size, analyses of data from wells in Central Pennsylvania were not performed.

Marcellus Shale produced water exhibits an Na:Cl ratio similar to that of other brines from Pennsylvania (Figure 5a). However, it differentiates itself from other brines by the concentration of divalent cations. Produced water from Marcellus Shale wells had slightly less Ca (Figure 5b), much less Mg (Figure 5c), and much more Sr (Figure 5d) than found in any other brines from PA. Although Ba data for produced water from all other formations are not available, indications are that the produced water from Marcellus Shale contains much more Ba compared to Lower Silurian and Upper Devonian formations. Furthermore, the produced water from Marcellus Shale does not exhibit the same trends in Ca:Cl and Mg:Cl ratios as other produced waters, especially during the early stages of the flowback period indicated by lower chloride concentrations. This behavior indicates that mixing with the formation brine cannot completely explain the salinity of the produced water over the entire life of a well.

The origin of the salinity in the produced water is better understood using ion concentrations that are plotted versus bromide concentration, as bromide in solution is normally conserved during water evaporation. (27) The conservative parameter MCl_2 is valuable when determining the chemical origin of chloride-rich brines. MCl_2 is calculated as follows:

$$MCl_2 \; (meq/L) = Ca^{2+} + Mg^{2+} + Ba^{2+} + Sr^{2+} - SO_4^{2-} - CO_3^{2-} \tag{1}$$

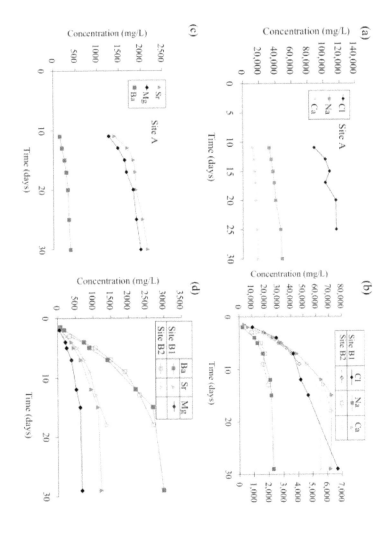

FIGURE 3: Concentration profiles of Na, Cl, Ca (a and b), and Mg, Sr, and Ba (c and d) for site A and sites B1 and B2, respectively. Day 0 corresponds to the end of the hydraulic fracturing process.

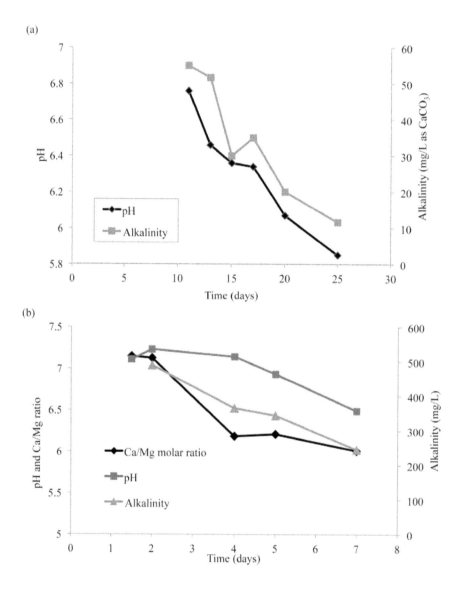

FIGURE 4: Variation of (a) pH and alkalinity with time for flowback water from site A and (b) Ca/Mg molar ratio, pH, and alkalinity with time for flowback water from site B1. Day 0 corresponds to the end of the hydraulic fracturing process.

MCl_2 is a conservative quantity during seawater evaporation (up to the point of $KMgCl_3$ precipitation). MCl_2 remains constant during precipitation of sulfates, carbonates, and NaCl. During seawater evaporation, the plot of $log[MCl_2] = f(log[Br])$ is a straight line of slope 1:1. It is represented by the following equation:(27)

$$log (MCl_2) = log[Br] + 0.011 \qquad\qquad (2)$$

Oil and gas brines from conventional reservoirs in SW Pennsylvania(26) follow the trend of seawater evaporation (Figure 6a) while the produced water from the Marcellus Shale shows an unusual relationship. High-salinity water samples closely match the seawater evaporation line, which suggests that the salinity of the produced water originates from the concentrated seawater. On the other hand, less concentrated produced water samples deviate from the expected relationship and exhibit either bromide enrichment or depletion of MCl_2 with respect to concentrated seawater. With time, produced water becomes concentrated in Ca, Mg, Ba, and Sr cations while sulfate and carbonate concentrations decrease. For the lowest salinity samples, the MCl_2 and bromide relationship deviates significantly from the 1:1 slope, indicating that the characteristics of these samples were influenced by the quality of the fracturing fluid and/ or chemical reactions that may be occurring in the formation during the contact with the shale.

The plot $log[Cl] = f(log[Br])$ for Marcellus Shale produced water samples exhibits a straight line with the slope that is very close to 1:1 for both Northeast and Southwest regions of Pennsylvania (Figure 6b). This finding suggests mixing of relatively dilute fracturing fluid with a brine concentrated beyond the point of halite saturation, which then exhibits a chloride:bromide ratio below that for seawater.(26) Finally, the logarithmic plot of [Cl] versus $[MCl_2]$ for Marcellus Shale produced water is a straight line with slope of approximately 0.7, which suggests that MCl_2 and Cl concentrations in flowback water are not governed by the simple dilution of the concentrated formation brine, as is the case for oil and gas brines from conventional reservoirs (Figure 7a). To better understand how flowback water composition evolves with time, the plot of [Cl] versus [MCl$_2$] was constructed for the data collected from wells A and B1 (Figure 7b). The average slope for both wells was again found to be around 0.7. With time, the flowback water salinity continues to increase, but is enriched in alkaline earths (or depleted in chloride) as compared to the trend expected for seawater evaporation. If the fracturing fluid were simply being mixed with the formation brine, the resulting plot of flowback water at different times would follow a straight line with the slope of 1:1. The slope exhibited by the flowback from these Marcellus Shale wells suggests that other mechanisms are involved and that the salinity of the early flowback water cannot be entirely explained by mixing of the fracturing fluid with existing formation water.

3.3.3 SPATIAL TRENDS IN FLOWBACK WATER CHEMISTRY IN PENNSYLVANIA

Although the flowback water chemistries from wells that were in close proximity exhibited similar trends (i.e., wells B1 and B2), it is difficult to predict composition of the flowback/produced water as it varies with time, location, and quality of the fracturing fluid. Nevertheless, it is important to estimate the concentrations of major cations of interest to reusing the flowback/produced water for hydraulic fracturing based on easily measurable water quality parameters. Therefore, all major cations included in this study were initially fitted to the following nonlinear regression model:

$$Y = \beta_0 X^{\beta_1} \varepsilon \tag{3}$$

where Y is the concentration of cation of interest, X is the chloride concentration, β_0 and β_1 are fitting constants, and ε is the error term.

The regression model was found to fit the data reasonably well with the data collected in SW being different from those collected in NE. However, the plots of residuals revealed nonconstant variance as well as nonsymmetrical distribution, suggesting that the model did not satisfy basic assumptions that the errors are normally distributed variables with zero mean and constant variance. To satisfy the normally distributed assumption and eliminate nonconstant variance, logarithmic transformations were performed as follows:

$$Y' = \log Y \qquad (4)$$

$$X' = \log[Cl] \qquad (5)$$

The resulting linear model

$$Y' = \log(\beta_0) + \beta_1 \times X' + \log(\varepsilon) \qquad (6)$$

was then tested and the residuals were normally distributed with mean equal to zero and constant variance indicating that the assumptions about error normality were correct.

Aside from chloride concentration, the intent was to determine if location is a significant regressor. Therefore, geographic location was incorporated in the linear model (6) as an indicator variable to identify if differences in composition between data from NE and SW were statistically significant. The following model was fitted to the data:

$$\log(Y) = C_1 + C_2 \times \log[Cl] + C_3 \times X_1 \times \log[Cl] + C_4 \times X_1 + \log(\varepsilon) \qquad (7)$$

where C_1, C_2, C_3, and C_4 are fitting constants, and $X_1 = 0$ if the observation is from SW and 1 if it is from NE.

Two equations were then obtained for the two geographical locations:

$$\log(Y_{SW}) = C_1 + C_2 * \log[Cl_{SW}] \qquad (8)$$

$$\log(Y_{NE}) = (C_1 + C_4) + (C_2 + C_3)*\log[Cl_{NE}]$$ (9)

Results of this analysis are summarized in Table 2. The significance of regression was tested to determine if at least one regression coefficient was different from zero. The null hypothesis was that all $C_i = 0$ against the alternative hypothesis that at least one C_i is different from zero. All tests were performed using alpha equal to 0.05. C_3 and C_4 are the constants that differentiate the NE data from the SW data. p-Values greater than 0.05 for these constants would mean that $C_3 = C_4 = 0$, and therefore no statistically significant difference exists between data from NE and SW. On the contrary, p-value below 0.05 for either or both C_3 and C_4 would indicate that the ion concentrations follow a different model for data from NE and SW.

TABLE 2: Fitting Constants for the Log-Log Multiple Regression Model[a]

	C1	p	C_2	p	C_3	p	C_4	p	R^2
sodium	0.176	0.012	0.888	<0.001	−0.024	0.257	0.097	0.331	0.983
calcium	−1.841	<0.001	1.195	<0.001	0.119	0.061	−0.687	0.020	0.936
magnesium	−2.692	<0.001	1.157	<0.001	0.147	0.132	−0.866	0.055	0.86
barium	−6.070	<0.001	1.761	<0.001	−0.389	0.125	3.107	0.009	0.744
strontium	−2.879	<0.001	1.254	<0.001	0.212	<0.001	−0.892	0.001	0.957
bromide	−2.299	<0.001	1.066	<0.001	−0.238	<0.001	0.976	<0.001	0.956

[a]*p-values in bold are greater than 0.05, indicating that the associated constant C_i is not a significant predictor.*

The p-values obtained for the significance of regression are infinitely small for all cases, meaning that at least one C_i is different from zero. In addition, high values for the coefficient of determination (R^2) indicate that the chloride concentration is a good predictor for the variations in concentrations of other ions. p-Values for C_3 and C_4 for sodium are particularly large, showing no statistical difference in the correlations with chloride concentrations between SW and NE. On the other hand, regressions for strontium and bromide reveal significant difference for the correlations between SW and NE with p-values for C_3 and C_4 much below 0.05. For barium and calcium, p-values for C_3 are greater than 0.05 while p-values for C_4 are lower than 0.05. The difference between SW and NE for these two ions relies on the multiplying constant but not the exponent of the power law. Magnesium, like sodium, does not exhibit statistically different behavior between SW and NE, but the p-value for C_4 is only slightly above the significance level. Overall, the concentrations of calcium, magnesium, and bromide are higher in the southwest part of Pennsylvania than in the northeast, while the opposite is true for barium and strontium. The opposite trend for strontium and calcium might indicate the transformation of strontium-containing aragonite into calcite through the precipitation of calcium ions and release of strontium ions.(28)

Among all the ions studied, barium in SW locations exhibited the lowest determination coefficient (51%) with particularly wide confidence intervals. Chloride concentration was not a sufficient predictor of the variation of barium. The results of the multiple regression displayed in Table 2 clearly indicated higher barium content for flowback water from the northeast part of the state, with concentrations reaching as high as 14 000 mg/L, while low concentrations were measured in the southwest even for samples containing high chloride concentrations. The geographic trend is illustrated by the heat map of Ba/Cl weight ratio shown in Figure 8. An average Ba/Cl ratio was calculated for each investigated county, and results are reported on a Pennsylvania map using the ArcGIS software. Counties located in the far northeast part of PA exhibited Ba/Cl ratio above 6%, while southwest counties had ratios up to 3%.

REFERENCES

1. Jaramillo, P.; Griffin, W. M.; Matthews, H. S.Comparative life-cycle air emissions of coal, domestic natural gas, LNG, and SNG for electricity generation Environ. Sci. Technol. 2007, 41, 6290– 6296

2. Burnham, A.; Han, J.; Clark, C. E.; Wang, M.; Dunn, J. B.; Palou-Rivera, I.Life-Cycle Greenhouse Gas Emissions of Shale Gas, Natural Gas, Coal, and Petroleum Environ. Sci. Technol. 2012, 46 (2) 619– 627

3. Howarth, R. W.; Santoro, R.; Ingraffea, A.Methane and the greenhouse-gas footprint of natural gas from shale formations Climatic Change 2011, 106 (4) 679– 690

4. Kargbo, D. M.; Wilhelm, R. G.; Campbell, D. J.Natural Gas Plays in the Marcellus Shale: Challenges and Potential Opportunities Environ. Sci. Technol. 2010, 44 (15) 5679– 5684

5. Milici, R. C.; Ryder, R. T.; Swezey, C. S.; Charpentier, R. R.; Cook, T. A.; Crovelli, R. A.; Klett, T. R.; Pollastro, R. M.; Schenk, C. J. USGS Assessment of Undiscovered Oil and Gas Resources of the Appalachian Basin Province; Fact Sheet 009-03; U.S. Geological Survey, 2002.

6. Engelder, T.Marcellus 2008: Report card on the breakout year for gas production in the Appalachian Basin. Fort Worth Basin Oil and Gas Magazine, August, 2009; pp 18– 22.

7. Chariag, B.Schlumberger, Inc. Maximize Reservoir Contact; Hart Energy Publishing, LP; Global Exploration and Production News, January 2007.

8. API Hydraulic Fracturing Operations-Well Construction and Integrity Guidelines; Report HF1; American Petroleum Institute, 2009.

9. King, G. E.Thirty years of gas shale fracturing: What have we learned?SPE Annual Technical Conference, Florence, Italy, September 19–22, 2010; Abstract SPE 133456.

10. Sirivedhin, T.; Dallbauman, L.Organic matrix in produced water from the Osage-Skiatook Petroleum Environmental Research site, Osage county, Oklahoma Chemosphere 2004, 57, 463– 469

11. Andrew, A. S.; Whitford, D. J.; Berry, M. D.; Barclay, S. A.; Giblin, A. M.Origin of salinity in produced waters from the Palm Valley gas field, Northern Territory Appl. Geochem. 2005, 20 (4) 727– 747

12. Orem, W. H.; Tatu, C. A.; Lerch, H. E.; Rice, C. A.; Bartos, T. T.; Bates, A. L.; Tewalt, S.; Corum, M. D.Organic compounds in produced waters from coalbed natural gas wells in the Powder River Basin, Wyoming, USA Appl. Geochem. 2007, 22 (10) 2240– 2256

13. Johnson, B. M.; Kanagy, L. E.; Rodgers, J. H.; Castle, J. W.Chemical, Physical, and Risk Characterization of Natural Gas Storage Produced Waters Water, Air, Soil Pollut. 2008, 191 (1–4) 33– 54

14. Fakhru'l-Razi, A.; Pendashteh, A.; Chuah Abdullah, L.; Radiah Awang Biak, D.; Siavash Madaeni, S.; Zainal Abidin, Z.Review of technologies for oil and gas produced water treatment J. Hazard. Mater. 2009, 170 (2–3) 530– 551

15. Alley, B.; Beebe, A.; Rodgers, J., Jr.; Castle, J. W.Chemical and physical characterization of produced waters from conventional and unconventional fossil fuel resources Chemosphere 2001, 85 (1) 74– 82

16. Chapman, E. C.; Capo, R. C.; Stewart, B. W.; Kirby, C. S.; Hammack, R. H.; Schroeder, K. T.; Edenborn, H. M.Geochemical and Strontium Isotope Characterization of Produced Waters from Marcellus Shale Natural Gas Extraction Environ. Sci. Technol. 2012, 46 (6) 3545– 3553

17. Gregory, K. B.; Vidic, R. D.; Dzombak, D. A.Water Management Challenges Associated with the Production of Shale Gas by Hydraulic Fracturing Elements 2011, 7 (3) 181– 186

18. APHA, AWWA, WPCF. Standard Methods for the Examination of Water and Drinking Water, 20th ed.; Washington, DC, 2000.

19. BOGM, Bureau of Oil and Gas Management. Frac and flowback water analytical data, inorganics, spreadsheet. Available at http://www.bfenvironmental.com/pdfs/ PADEP_Frac_Flow_Back_Water_Study__Presence_of_Inorganics.pdf, last access 06/20/ 2012.

20. Hayes, T. Sampling and Analysis of Water Streams Associated with the Development of Marcellus Shale Gas; Final report prepared for the Marcellus Shale Coalition, December 31, 2009.

21. Barbot, E.; Vidic, R.Potential for abandoned mine drainage as water supply for hydraulic fracturing in the Marcellus Shale. 244th ACS National Meeting & Exposition, August 19–23, 2012, Philadelphia, Pennsylvania.

22. Osborn, S. G.; McIntosh, J. C.Chemical and isotopic tracers of the contribution of microbial gas in Devonian organic-rich shales and reservoir sandstones, northern Appalachian Basin Appl. Geochem. 2010, 25 (3) 456– 471

23. Roen, J. B.Geology of the Devonian black shales of the Appalachian Basin Org. Geochem. 1984, 5 (4) 241– 254

24. Boyce, M. L.Sub-surface stratigraphy and petrophysical analysis of the Middle Devonian interval of the central Appalachian basin; West Virginia and southwest Pennsylvania. Thesis. West Virginia University, 2010, 159 pp.

25. Blauch, M. E.; Myers, R. R.; Lipinski, B. A.; Houston, N. A.Marcellus Shale postfrac flowback waters – Where is all the salt coming from and what are the implications?; SPE 125740; Society of Petroleum Engineers, 2009.

26. Dresel, P. E.; Rose, A. W. Chemistry and Origin of Oil and Gas Well Brines in Western Pennsylvania; 4th ser.; Pennsylvania Geological Survey, 2010; Open-File Report OFOG 10-01.0, 48 pp.

27. Carpenter, A. B.Origin and chemical evolution of brines in sedimentary basins Okla. Geol. Surv., Circ. 1978, 79, 60– 76[CAS]

28. Katz, A.; Sass, E.; Starinsky, A.; Holland, H. D.Strontium behavior in the aragonite-calcite transformation: An experimental study at 40–98 °C Geochim. Cosmochim. Acta 1972, 36 (4) 481– 496

There are several figures that are not available in this version of the article. To view this additional information, please use the citation on the first page of this chapter.

PART II

POTENTIAL ENVIRONMENTAL EFFECTS OF FRACKING WASTEWATER

CHAPTER 4

Shale Gas Development Impacts on Surface Water Quality in Pennsylvania

SHEILA M. OLMSTEAD, LUCIJA A. MUEHLENBACHS, JHIH-SHYANG SHIH, ZIYAN CHU, and ALAN J. KRUPNICK

With the advance of hydraulic fracturing technology and improvements in horizontal well drilling, the development of natural gas supplies from deep shale formations has expanded and US natural gas supply estimates have risen dramatically (1). These resources have significant economic value and could generate local air quality benefits if gas displaces coal in electricity generation and for climate change if fugitive methane emissions are sufficiently small (2). Nonetheless, shale gas development has drawn significant public and regulatory attention to potential negative environmental externalities, particularly water quality impacts in the Marcellus Shale region (3, 4).

Groundwater impacts of shale gas development have been considered in the literature. Methane may migrate from shale gas wells into drinking water wells in Pennsylvania and New York (5). Shale formation brine may

Reprinted with permission from the authors. Shale Gas Development Impacts on Surface Water Quality in Pennsylvania. © Olmstead SM, Muehlenbachs LA, Shih J-S, Chu Z, and Krupnick AJ. Proceedings of the National Academy of Sciences of the United States of America *110,13 (2013). doi: 10.1073/pnas.1213871110.*

also naturally migrate to groundwater aquifers in Pennsylvania, although this result is debated in the literature (6) and no association has been found with the location of shale gas wells (7). Case studies of isolated incidents of groundwater contamination also suggest links with shale gas activity (8).

The potential risk to New York City's surface water supply from the Delaware River Basin was a primary driver behind a 2011 ban on hydraulic fracturing in New York State. In contrast to the case of groundwater, however, empirical estimates of the effects of shale gas development on surface water quality are not available, although the issue has been raised in the recent literature (9, 10).

We conduct a large-scale statistical examination of the extent to which shale gas development affects surface water quality. Focusing on the Marcellus Shale, a major US shale play, we construct a Geographic Information Systems (GIS) database from several publicly available sources, including 20,283 water quality observations in Pennsylvania (2000–2011), shale gas well locations, shipments of shale gas waste to treatment facilities, and water body characteristics. We exploit temporal and spatial variation in the location of wells and waste treatment facilities relative to water quality monitors to identify impacts on downstream water quality. Using regression analysis, we find measurable impacts of upstream shale gas activity on downstream water quality. Increasing the upstream density of wastewater treatment plants that release treated shale gas waste to surface water by 1 SD increases downstream chloride (Cl^-) concentrations by 10–11%. A 1-SD increase in the density of well pads upstream increases total suspended solid (TSS) concentrations downstream by 5%. In contrast, we find no statistically significant impact of wells on downstream Cl^- concentrations or of waste treatment on downstream TSS concentrations. These findings are consistent with concerns raised for surface water quality in the literature: Shale gas wastewater is typically high in Cl^- (among other dissolved solids), making it difficult to treat, and the construction of well pads, pipelines, and roads can increase sediment runoff and TSS (9, 10).

The econometric approach used here cannot identify or rule out individual instances of water quality contamination. The analysis models average impacts at coarse temporal and spatial scales as a function of shale gas development, controlling for other factors. Thus, it is a complement to physical science studies that would make precise connections between water quality changes at a fine temporal and spatial scale and specific shale gas activities.

4.1 SHALE GAS ACTIVITY AND WATER QUALITY INDICATORS

Indicators of water quality impacts from shale gas development must meet three criteria for the analysis: They are associated with shale gas development, they are observed at a large number of water quality monitors and with enough spatial and temporal variation relative to shale gas activity to support statistical analysis, and they have the potential to cause water quality damage. Concentrations of Cl⁻ and TSS meet these criteria.

Brine from conventional oil and gas operations has been associated with increased stream Cl⁻ levels in other regions (11, 12). Shale gas development generates large quantities of flowback and produced water high in Cl⁻ (9). The peak in fluid storage and transport occurs during well fracturing and completion, when 2–4 million gallons of freshwater and fracturing fluids transported by truck or (for freshwater) pipeline are pumped into a well (13). From 10–70% of this volume may return as flowback, along with formation brine and naturally occurring contaminants, such as heavy metals and radionuclides. Direct discharge of fluids from well sites is illegal. Fluids are collected at well pads and transported on- and off-site for reuse, recycling, treatment, and disposal. Media coverage of Marcellus Shale development suggests potential risks of leakage from storage pits and impoundments, spills, and other accidental releases (3, 14).

A large and increasing fraction of shale gas wastewater in Pennsylvania is recycled for use in other well completions. Some waste is shipped to deep injection wells in Ohio and other neighboring states. (Most of Pennsylvania is geologically unsuitable for deep injection wells. Seismic concerns have arisen regarding injection of large quantities of shale gas waste into these wells (15, 16).) Operators have also shipped shale gas waste to municipal and industrial wastewater treatment facilities, although the effectiveness of treatment processes in these facilities for removing contaminants in shale gas waste is poorly understood, and salts (including Cl⁻) are of particular concern (17). (Shipments to municipal facilities stopped as of January 2012; those to industrial facilities continue.) Average total dissolved solid (TDS) concentrations in shale gas waste range from 800 to 300,000 mg/L, typical ocean water concentration is 35,000 mg/L, and freshwater concentration is 100–500 mg/L (18). Concern over the limited capacity of Pennsylvania rivers and streams to assimilate TDSs that remain in wastewater treatment plant effluent considering existing sources, such as coal mine drainage and conventional wastewater effluent, prompted the introduction of new state wastewater treatment standards for TDSs in 2011. [In 2008, monitors detected record TDS levels (mainly Cl⁻ and sulfates) in sections of the Monongahela River during low late summer-early fall flows, and 13 public drinking water system intakes in Pennsylvania and West Virginia exceeded secondary maximum contaminant levels under the Safe Drinking Water Act, persisting through December 2008 (18).]

Elevated or fluctuating Cl⁻ concentrations can directly damage aquatic ecosystems (19). Cl⁻ may also mobilize heavy metals, phosphates, and other chemicals present in sediment (20). Treatment of waste high in Cl⁻ is expensive because the Cl− is not easily removed by chemical or biological processes once it is in solution (21); thus, high Cl⁻ concentrations may also increase costs for downstream water users (e.g., industrial or drinking water facilities).

Land clearing and construction can increase TSS in local water bodies, particularly when precipitation accelerates sediment transport, increases flow rates so that water carries more and larger sediments, and resuspends sediments. Pad construction, changes to local roadways, pipeline construction, and other shale gas development activities could contribute to this problem (9). Gas well sites in Texas have been shown to produce sediment loads comparable to traditional construction sites (22). However, the US Energy Policy Act of 2005 generally exempts oil and gas construction sites from Clean Water Act (CWA) stormwater regulations. In Pennsylvania, non–oil-and-gas construction sites larger than 1 acre must install erosion and sediment control infrastructure; shale gas sites larger than 5 acres must file erosion and sediment control plans. Most shale gas well sites are not large enough to trigger this review, although many Marcellus Shale operators do install stormwater control infrastructure (13). In addition to impacts from infrastructure, TSS concentrations could be increased by shale gas waste treatment, although most wastewater treatment plants are designed to remove suspended solids (13).

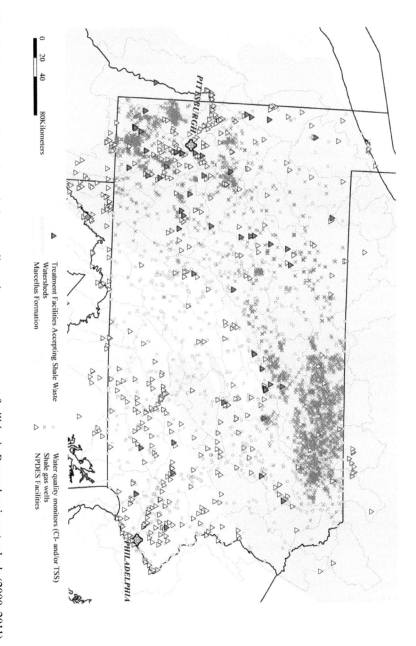

FIGURE 1: Surface water quality monitors, shale gas wells, and wastewater treatment facilities in Pennsylvania watersheds (2000–2011).

TSS (silt, decaying organic matter, industrial wastes, and sewage that can be trapped by a fine filter) in surface water reduce available sunlight, raise temperature, decrease dissolved oxygen and clarity, and ultimately damage biological condition (23). Solids can also clog or scour pipes and machinery for downstream water users, increasing costs.

4.2 DATA

Our GIS database combines several sources of data (additional details are provided in SI Data). The Storage and Retrieval Data Warehouse (STORET) database of the Environmental Protection Agency (EPA) provided 8,364 Cl⁻ concentration observations from 860 Pennsylvania water quality monitors between January 2000 and December 2011 (with 98 d, on average, between readings at a monitor). TSS concentrations from the STORET database for the same time period comprise 11,919 observations from 644 monitors (with readings every 55 d, on average).

The independent variables of greatest interest capture the density of shale gas wells in a monitor's watershed and the scale of shale gas waste treatment and release in a monitor's watershed. The latitude and longitude and the drilling and completion dates of 4,908 shale gas wells drilled through December 2011 were obtained from the Pennsylvania Department of Environmental Protection (PADEP) and the Pennsylvania Department of Conservation and Natural Resources (PADCNR). The PADEP also provided the destination and quantity of wastewater shipments from wells to 74 permitted treatment facilities that accepted shale gas waste at some point between 2004 (the year of the first observed shipment) and 2011. In 2004, 0.003 million barrels (MMbbl) of liquid shale gas waste were shipped to 3 treatment facilities; in 2011, about 17.7 MMbbl were shipped to 50 facilities. We obtained the latitude and longitude of these facilities, as well as others permitted under the CWA's National Pollutant Discharge Elimination System (NPDES), from the EPA. Daily precipitation data were downloaded from the National Oceanic and Atmospheric Administration's National Climatic Data Center.

We generated well counts, shale gas waste treatment facility counts, and waste shipment amounts in the upstream portion of each monitor's watershed. Thus, we focus only on within-watershed impacts, excluding any potential impacts further downstream. (Estimating impacts in downstream watersheds would require hydrological modeling of the attenuation over time and space of contaminant concentrations, which is beyond the scope of this study. However, the most significant impacts, if any, should be detectable within watersheds.) Watershed boundaries were determined using a US Geological Survey digital elevation model, HYDRO1k. Flow direction and accumulation were determined using a 1-km by 1-km grid size in HYDRO1k in ArcGIS. A flow accumulation threshold of 1,000 km^2 was used to delineate 79 Pennsylvania watersheds, averaging 1,502 km^2.

4.3 RESEARCH DESIGN

The research design exploits spatial variation in the location of water quality monitors, shale gas wells, and NPDES-permitted waste treatment facilities that have accepted shale gas waste (Fig. 1), along with intertemporal variation generated by the timing of well development and waste shipments. The effects of interest are estimated using regression analysis (Supporting Information).

We test for the impact of shale gas development on surface water Cl$^-$ concentrations via two potential pathways. First, to consider the possibility of accidental releases from well sites, we analyze the impact on Cl$^-$ concentrations of the density of shale gas wells upstream in a water quality monitor's watershed (the count of upstream wells on the date a sample was drawn, divided by the area of the upstream portion of the watershed). The statistical models (Eq. S1) also test whether any estimated impacts of wells on downstream Cl$^-$ are more significant during well fracturing and completion. Second, the models examine the impact of waste fluid shipments on downstream Cl$^-$ concentrations, considering both the density of treatment facilities accepting shale gas waste (the count of upstream facilities accepting waste during the period in which the sample is drawn, divided by upstream area) and the quantity of shipments reported to the PADEP upstream in a water quality monitor's watershed. We also test for differences in the impacts of upstream treatment of shale gas waste by different types of treatment facilities. Facilities accepting Pennsylvania shale gas waste between 2004 and 2011 include municipal sewage treatment plants, or publicly owned treatment works (POTWs), and industrial centralized waste treatment (CWT) facilities. Facilities' NPDES permits determine their eligibility to treat shale gas waste, and regulation of these flows changed during the period of observation. Regulatory changes in January 2011 increased the stringency of treatment requirements, although a specific group of grandfathered POTWs and CWTs was initially exempt from new requirements and then subject to a voluntary ban on waste shipments from operators in May 2011 (Supporting Information). The statistical models test for differential effects on downstream Cl^- concentrations of shale gas waste treatment at facilities affected by these regulatory changes and those that were not.

We analyze the impacts of shale gas development on surface water TSS concentrations via the same two basic pathways considered for Cl^-: impacts from wells and impacts from waste disposal at permitted treatment facilities. However, rather than examining more closely the period during well completion, as we did for Cl^-, the TSS models consider whether any estimated impacts of wells on TSS are concentrated between the permit date and spud date (the date on which well drilling begins), when land clearing and pad construction would take place.

To identify the impacts of shale gas wells and waste treatment on Cl^- and TSS concentrations, we must adequately control for other contributing factors. All models include controls for precipitation, summing daily precipitation in a monitor's watershed on the day of a monitor observation plus cumulative precipitation on the 3 d before each observation. The models control for other potential influences on Cl^- and TSS concentrations using a standard econometric approach, including groups of fixed effects (FEs) as controls. We control for average concentrations at each monitor over time with monitor FEs. Year-month FEs (132 for the months January 2000–December 2011) control for all time-varying characteristics of Pennsylvania water quality, generally, that are correlated over space (e.g., trends in economic activity, such as those related to the recent recession). Watershed-calendar-month (January–December) FEs control for seasonal changes in contaminant concentrations that may be specific to a watershed.

TABLE 1: Estimated impacts of shale gas wells and waste treatment on downstream Cl− concentrations (milligrams per liter)

Variable	(1)	(2)	(3)	(4)	(5)	(6)
Cumulative precipitation (4 d), mm	−0.003***	−0.003***	−0.003***	−0.003***	−0.003***	−0.003***
	(0.001)	(0.001)	(0.001)	(0.001)	(0.001)	(0.001)
Gas wells upstream/km²		17.076		10.312		20.204
	(16.219)	(16.952)		(19.245)		
Gas wells upstream (0–90 d)/km²			55.196			
			(126.292)			
Gas wells upstream (90–180 d)/km²			−36.500			
			(93.82)			
Facilities accepting waste upstream/km²		2,223.660***	2,240.228***	2,116.823***	2,086.356***	
		(681.868)	(696.602)	(696.135)	(683.888)	
Waste quantity treated upstream, MMbbl/km²				152.873	205.601*	
				(140.226)	(119.618)	
Nonaffected facilities accepting waste/km²						2,625.909
						(4,164.453)
Affected facilities accepting waste/km²						1,927.716**
						(902.173)
N	8,402	8,364	8,364	8,364	8,364	8,364
Mean Cl−, mg/L	19.077	19.074	19.074	19.074	19.074	19.074
R²	0.499	0.499	0.499	0.499	0.499	0.499

Sample includes all monitor measurements of Cl− (milligrams per liter), 2000–2011. Variables divided by square kilometers are divided by the area of the watershed that is upstream of the monitor. All regressions include FEs for year-month, watershed-calendar month, and monitor. Reported SEs are robust and clustered by watershed. Statistically significant at the *10% level; ** 5% level; ***1% level.

Together, the FEs narrow the sources of spatial and intertemporal variation in contaminant concentrations from which the effects of interest are identified, so as to exclude potential confounders. For example, an important source of Cl⁻ in Pennsylvania rivers and streams is road salt. The monitor FEs control for the fact that some monitors consistently receive more road salt from upstream sources than others. The year-month FEs control for the fact that for all Pennsylvania monitors, some months between January 2000 and December 2011 witness higher Cl⁻ concentration from road salt than others, and the watershed-calendar-month FEs control for the fact that some months of the year are typically subject to higher concentrations of Cl⁻ from road salt than others, and the magnitude of this interannual variation varies by watershed. Further analysis decomposing some of these FEs, to demonstrate that the models control sufficiently for important potential confounders, is provided in Supporting Information.

4.4 RESULTS

4.4.1 ESTIMATED IMPACTS ON CL□ CONCENTRATIONS.

A FE regression of observed Cl⁻ concentration on the FEs and precipitation shows that an increase in precipitation reduces Cl⁻ concentration. The magnitude of this simple dilution result does not change across specifications (Table 1).

The density of shale gas wells upstream in a monitor's watershed has a statistically insignificant effect on Cl⁻ concentration downstream, although coefficient estimates are positive (Table 1, columns 2, 4, and 6). Limiting the well density variable to only those wells spudded within 0–90 d and within 90–180 d before a Cl⁻ sample is drawn at a downstream water quality monitor should capture the greatest potential for accidental releases affecting surface water, because reported completion dates in our dataset are, on average, 80 d after a well's spud date (SI Data). (We cannot construct variables centered on a known completion date, because firms report completion dates to the PADEP and/or

PADCNR for only 1,815 of the 4,908 wells drilled through December 2011. A nonreported completion date does not necessarily indicate that a well has not been completed; 877 wells without a completion date report production before the end of 2011.) Nonetheless, we find no statistically significant impact of wells at this development stage on observed Cl⁻ concentrations (Table 1, column 3).

In contrast, the density of waste treatment facilities accepting shale gas waste upstream in a monitor's watershed increases Cl⁻ concentrations at monitors downstream (Table 1, columns 2–5). The annual quantity of wastewater shipped to waste treatment facilities upstream of a monitor, divided by the upstream area of the watershed, has varying statistical significance (Table 1, columns 4–5). Waste shipment quantity data are potentially problematic (SI Data). However, they represent the only data currently available on the quantity of shale gas waste treated and released by particular facilities. The quantity of waste shipped to treatment facilities in a watershed is correlated with the density of shale gas wells. If the density of wells is included in the analysis, the waste quantity is positive but statistically insignificant (Table 1, column 4). If the wells variable is dropped, the waste quantity variable is positive and weakly significant (Table 1, column 5). Because the coefficients of the wells are statistically insignificant in all models, this suggests a potential marginal effect of the quantity of waste treated and released upstream on Cl⁻ concentrations, controlling for the spatial density of treating facilities.

The impact of treated shale gas waste on downstream Cl⁻ concentrations may vary across treatment facilities. We divide facilities into those that were affected by the 2011 voluntary request to suspend shipments (POTWs and grandfathered CWTs) and those that were not affected (non-grandfathered CWTs). Results suggest that the observed increase in downstream Cl⁻ concentrations is due more to the facilities that have received regulatory attention than to those that have not; the coefficient on the affected facilities is positive and statistically significant (Table 1, column 6). However, the PADEP data indicate that some affected facilities accepted shale gas waste during the period July 2011–December 2011, although operators were to have voluntarily stopped sending waste to these facilities in May 2011 (Supporting Information).

4.4.2 ESTIMATED IMPACTS ON TSS CONCENTRATIONS.

A simple model regressing observed TSS concentrations on the FEs and precipitation finds that precipitation in the watershed increases TSS concentrations (Table 2). The magnitude of this expected effect does not change as we alter model specification.

TABLE 2: Estimated impacts of shale gas wells and waste treatment on downstream TSS concentrations (milligrams per liter)

Variable	(1)	(2)	(3)	(4)	(5)
Cumulative precipitation (4 d), mm	0.086*	0.086***	0.086***	0.085***	0.086***
	(0.025)	(0.025)	(0.025)	(0.026)	(0.025)
Facilities accepting waste upstream/km^2		706.063	692.274	691.812	695.183
		(454.138)	(456.297)	(461.978)	(451.551)
Gas wells upstream/km^2		45.965**			
		(22.867)			
Well pads upstream/km^2			97.072*	3.427	100.921
			(54.548)	(127.972)	(71.640)
Well pads upstream/km^2 × (4-d precipitation)				0.514	
				(0.867)	
Well pads permitted, prespud upstream/km^2					−102.865
					(892.745)
N	11,919	11,919	11,919	11,919	11,919
Mean TSS, mg/L	20.392	20.392	20.392	20.392	20.392
R^2	0.283	0.283	0.283	0.283	0.283

*Sample includes all monitor measurements of TSS (milligrams per liter), 2000–2011. Variables divided by square kilometers are divided by the area of the watershed that is upstream of the monitor. All regressions include FEs for year-month, watershed-calendar month, and monitor. Reported SEs are robust and clustered by watershed. Statistically significant at the *10% level; ** 5% level; ***1% level.*

The density of upstream waste treatment facilities accepting shale gas waste has a statistically insignificant effect on downstream TSS concentrations, although the coefficients are positive (Table 2, columns 2–5). The density of shale gas wells upstream in a monitor's watershed, in contrast, has a positive and significant impact on downstream TSS concentrations (Table 2, column 2).

We also test for the relative impact of well pad preparation vs. activities specific to individual wells. The data do not indicate which wells are on the same well pad; thus, we assume that wells within 1 acre share a pad. The average number of wells on a pad using this method is about 3.7.* We then create a variable describing the density of well pads upstream in a monitor's watershed. If pad preparation is a more significant driver of TSS impacts than individual wells, the magnitude of the well pad estimate should be less than the impact of an individual well, multiplied by the average number of wells on a pad. The well pad variable is positive, weakly significant (Table 2, column 3), and about twice the magnitude of the variable for individual wells (Table 2, column 2). Although we may have introduced some attenuation bias from measurement error in constructing the well pad variable, this result is consistent with the hypothesis that the net TSS impacts of wells may be due more to well pad preparation than to activities associated with individual wellbores.

If the TSS impacts we estimate from shale gas wells are due to stormwater-related transport of sediment to water bodies, they should intensify with rain. Interacting the well pad density variable with precipitation does not support this hypothesis; coefficients on well pad density and the interaction term are positive but statistically insignificant (Table 2, column 4). Finally, if these impacts are related primarily to disturbance from land clearing and pad construction, we should observe stronger impacts from pads on which the first observed well is between the permit date and the spud date than from pads that are further along in development. However, permitted but undrilled wells are negative and insignificant, and the statistical significance of the well pads variable disappears as well (Table 2, column 5). (The coefficient estimate on the density of well pads for which one or more wells are permitted but not yet spudded remains insignificant if we drop the variable capturing the total number of well pads.)

4.5 INTERPRETATION AND DISCUSSION

Results for Cl⁻ suggest that the presence of shale gas wells upstream in a monitor's watershed does not raise observed concentrations but that the treatment and release of wastewater from shale gas wells by permitted facilities upstream in a monitor's watershed does. These results are not consistent with the presence of significant flows of high-Cl⁻ shale gas waste through accidental releases directly into surface water from well sites. However, surface water disposal of treated waste from shale gas wells represents a potentially important water quality burden. Taking into account average watershed size and mean Cl⁻ concentrations, the coefficients in Table 1 suggest that a 1-SD increase in the spatial density of upstream waste treatment facilities (an additional 1.5 facilities treating waste upstream in a watershed) results in a 10–11% increase in Cl⁻ downstream, depending on the specification. (We calculate this effect as follows: (SD of upstream facilities per km² × coefficient)/average Cl⁻ concentration.) Shale gas wastewater shipments to Pennsylvania POTWs have ceased as of January 2012. However, the documentation of a measurable surface water impact from only the first years of burgeoning development is relevant to CWTs that continue to treat shale gas wastewater as well as to other jurisdictions considering treatment by POTWs. Furthermore, apart from Cl⁻, many other wastewater constituents could potentially reach surface water, although available data on their concentrations is limited (Supporting Information).

Results for TSS suggest a different pathway of potential concern. The presence of shale gas wells upstream in a monitor's watershed raises observed TSS concentrations downstream. Increasing the average density of well pads upstream in a monitor's watershed by 1 SD (an additional 18 well pads) results in about a 5% increase in observed TSS concentrations (Table 2, column 3). Shale gas waste shipments to permitted treatment facilities do not appear to raise TSS concentrations. In the case of TSS, the primary water quality burden may be associated with the process of clearing land for infrastructure. However, given that we do not detect an increase in TSS impacts of well pads during precipitation events, or an increase associated with well pads in construction, the particular mechanisms through which shale gas infrastructure may increase TSS in local water bodies are unclear. Further analysis using data on pipeline and new road construction would be helpful in this regard. The observed increase in TSS concentrations could potentially be associated with spills or other emissions at well sites, rather than construction, but the inability of our models to detect increases in Cl⁻ from well sites (a strong marker for shale gas waste, relative to TSS, which have many more sources) is not consistent with this possibility.

The nature of surface water contamination from shale gas development considered here is qualitatively different from the groundwater concerns explored in the literature (5, 7). Although groundwater concerns may have primarily to do with contamination directly from wellbores or shale formations, surface water concerns may have primarily to do with off-site waste treatment and aboveground land management.

The effects of shale gas development on surface water quality that we estimate control for average contaminant concentrations at a monitor, average concentrations for each month in the data, and average concentrations in each calendar month by watershed, as well as precipitation. Thus, the shale gas coefficient estimates are identified from within-monitor variation in contaminant concentrations, controlling carefully for exogenous trends in water quality. However, this approach cannot confirm or rule out individual accidental releases of flowback and other fluids to surface water. Our approach is complementary to physical science studies that would establish the exact mechanisms through which shale gas development may affect downstream surface water quality.

Finally, the economically optimal level of pollution is generally not zero, and further work would be necessary to quantify the benefits and costs of the shale gas development activities generating these externalities, including monetization of our estimated Cl⁻ and TSS impacts. The results highlight the need for further research on the surface water quality impacts of shale gas development, and they may provide input to operator decisions and regulatory processes regarding well location, waste disposal, erosion control, and contaminant monitoring.

REFERENCES

1. US Energy Information Administration (2012) Annual Energy Outlook 2012, DOE/ EIA-0383(2012) (US Energy Information Administration, Washington, DC).
2. Alvarez RA, Pacala SW, Winebrake JJ, Chameides WL, Hamburg SP (2012) Greater focus needed on methane leakage from natural gas infrastructure. Proc Natl Acad Sci USA 109(17):6435–6440.
3. Urbina I, (February 23–December 31, 2011) Drilling down series. The New York Times. Available at http://www.nytimes.com/interactive/us/DRILLING_DOWN_ SERIES.html. Accessed February 21, 2013.
4. US Environmental Protection Agency, Office of Research and Development, (2011) Plan to Study the Potential Impacts of Hydraulic Fracturing on Drinking Water Resources EPA/600/R-11/122 (US Environmental Protection Agency, Washington, DC).
5. Osborn SG, Vengosh A, Warner NR, Jackson RB (2011) Methane contamination of drinking water accompanying gas-well drilling and hydraulic fracturing. Proc Natl Acad Sci USA 108(20):8172–8176.
6. Saiers JE, Barth E (2012) Potential contaminant pathways from hydraulically fractured shale aquifers. Ground Water 50(6):826–828, discussion 828–830.
7. Warner NR, et al. (2012) Geochemical evidence for possible natural migration of Marcellus Formation brine to shallow aquifers in Pennsylvania. Proc Natl Acad Sci USA 109(30):11961–11966.
8. US Environmental Protection Agency, Office of Research and Development, National Risk Management Research Laboratory, (2011) Investigation of Ground Water Contamination near Pavillion, Wyoming, EPA 600/R-00/000 (US Environmental Protection Agency, Ada, OK).
9. Entrekin S, Evans-White M, Johnson B, Hagenbuch E (2011) Rapid expansion of natural gas development poses a threat to surface waters. Front Ecol Environ 9(9):503–511.
10. Rozell D, Reaven S (2012) Water pollution risk associated with natural gas extraction from the Marcellus Shale. Risk Anal 32(8):1382–1393.
11. Nance H (2006) Tracking salinity sources to Texas streams: Examples from West Texas and the Texas Gulf Coastal Plain. Gulf Coast Assoc Geol Soc Trans 56:675–693.
12. Shipley F (1991) Oil field-produced brines in a coastal stream: Water quality and fish community recovery following long term impacts. Tex J Sci 43(1):51–64.
13. Veil J (2010) Water Management Technologies Used by Marcellus Shale Gas Producers, Final Report, DOE Award No. FWP 49462 (US Department of Energy, Argonne National Laboratory, Argonne, IL).
14. Slater D (2011) Watershed moment. Sierra 96(5):14.
15. Ohio Department of Natural Resources (2012) Preliminary Report on the Northstar 1 Class II Injection Well and the Seismic Events in the Youngstown, Ohio Area (Ohio Department of Natural Resources, Columbus, OH) Available at http://ohiodnr. com/downloads/northstar/UICreport.pdf. Accessed February 21, 2013.

16. Ellsworth W, et al. (2012) Are seismicity rate changes in the midcontinent natural or manmade? Seismol Res Lett 83(2):403.

17. Soeder D, Kappel W (2009) Water Resources and Natural Gas Production from the Marcellus Shale, U.S. Geological Survey Fact Sheet 2009–3032 (US Geological Survey, West Trenton, NJ). Available at http://pubs.usgs.gov/fs/2009/3032/pdf/FS2009-3032.pdf. Accessed February 21, 2013.

18. Pennsylvania State University, College of Agricultural Sciences (2010) Shaping Proposed Changes to Pennsylvania's Total Dissolved Solids Standard: A Guide to the Proposal and the Commenting Process (Pennsylvania State Univ, University Park, PA) Available at http://extension.psu.edu/water/conservation/consumption-and-usage/TDS-highres-updateDec09.pdf/view. Accessed February 21, 2013.

19. Kaushal SS, et al. (2005) Increased salinization of fresh water in the northeastern United States. Proc Natl Acad Sci USA 102(38):13517–13520.

20. Nelson S, Yonge D, Barber M (2009) Effects of road salts on heavy metal mobility in two eastern Washington soils. J Environ Eng 135(7):505–510.

21. Novotny E, Stefan H (2010) Projections of chloride concentrations in urban lakes receiving road de-icing salt. Water Air Soil Pollut 211(1-4):261–271.

22. Williams H, Havens D, Banks K, Wachal D (2008) Field-based monitoring of sediment runoff from natural gas well sites in Denton County, Texas, USA. Environ Geol 55(7):1463–1471.

23. US Environmental Protection Agency, Office of Research and Development and Office of Water, (2006) Wadeable Stream Assessment: A Collaborative Survey of the Nation's Streams, EPA 841-B-06-002 (US Environmental Protection Agency, Washington, DC).

24. New York Department of Environmental Conservation (2011) Revised draft supplemental generic environmental impact statement on the oil, gas and solution mining regulatory program, well permit issuance for horizontal drilling and high-volume hydraulic fracturing to develop the Marcellus Shale and other low-permeability gas reservoirs. Available at www.dec.ny.gov/docs/materials_minerals_pdf/rdsgeisexecsum0911.pdf. Accessed February 21, 2013.

There are several supplemental files that are not available in this version of the article. To view this additional information, please use the citation on the first page of this chapter.

CHAPTER 5

Geochemical and Isotopic Variations in Shallow Groundwater in Areas of the Fayetteville Shale Development, North-Central Arkansas

NATHANIEL R. WARNER, TIMOTHY M. KRESSE,
PHILLIP D. HAYS, ADRIAN DOWN, JONATHAN D. KARR,
ROBERT B. JACKSON, AND AVNER VENGOSH

5.1 INTRODUCTION

The combined technological development of horizontal drilling and hydraulic fracturing has enabled the extraction of hydrocarbons from unconventional sources, such as organic-rich shales, and is reshaping the energy landscape of the USA (Kargbo et al., 2010 and Kerr, 2010). Unconventional natural gas currently supplies ~20% of US domestic gas production and is projected to provide ~50% by 2035 (USEIA, 2010). Therefore, ensuring that unconventional natural gas resource development results in the minimal possible negative environmental impacts is vital, not only for domestic production within the USA, but also for establishing guidance for worldwide development of shale gas resources. Recent work in the

Geochemical and Isotopic Variations in Shallow Groundwater in Areas of the Fayetteville Shale Development, North-Central Arkansas. © *Warner NR, Kresse TM, Hays PD, Down A, Karr JD, Jackson RB, and Vengosh A.* Applied Geochemistry **35** *(2013). Licensed under a Creative Commons Attribution 3.0 Unported License, http://creativecommons.org/licenses/by/3.0/.*

Marcellus Shale basin demonstrated a relationship between CH_4 concentrations in shallow groundwater and proximity of drinking water wells to shale-gas drilling sites in northeastern Pennsylvania, suggesting contamination of shallow groundwater by stray gas (Osborn et al., 2011a). In addition, a previous study has shown evidence for natural pathways from deep formations to shallow aquifers in northeastern Pennsylvania that may allow leakage of gas or brine, and might pose a potential threat to groundwater in areas of shale gas extraction (Warner et al., 2012). While previous studies have focused on the Pennsylvania and New York portion of the northern Appalachian Basin, many other shale-gas basins currently are being developed that have not been examined for potential effects on water quality. One of the critical aspects of potential contamination of shallow aquifers in areas with shale-gas development is the hydraulic connectivity between shale and other deep formations and overlying shallow drinking water aquifers. Here the quality and geochemistry of shallow groundwater directly overlying the Fayetteville Shale (FS) in north-central Arkansas is investigated. The Fayetteville Shale is an unconventional natural gas reservoir with an estimated total production of 906 billion m^3 (USEIA, 2011). Since 2004, approximately 4000 shale-gas wells have been drilled there, including both vertical wells and, more recently, horizontal wells.

In this study, water samples from 127 shallow domestic wells in the Hale, Bloyd and Atoka Formations in north-central Arkansas and six flowback/produced water samples from the underlying FS were analyzed in an attempt to identify possible groundwater contamination. Five of the produced water samples were collected within 21 days of fracturing (i.e., defined as flowback water) and a single sample was collected at about a year following hydraulic fracturing (i.e., defined as produced water). The concentrations of major anions (Cl, SO_4, NO_3, Br, and dissolved inorganic C [DIC]), cations (Na, Ca, Mg and Sr), trace elements (Li and B), and for a smaller subset of samples dissolved CH_4 and selected isotopic tracers ($\delta^{11}B$, $^{87}Sr/^{86}Sr$, δ^2H, $\delta^{18}O$, $\delta^{13}C_{DIC}$, and $\delta^{13}C_{CH4}$) were determined. Using multiple geochemical and isotopic tracers together with their geospatial distribution provides a multidimensional approach to examine potential groundwater contamination in areas of shale gas development. It is hypothesized that shallow groundwater could be contaminated by stray gas migration, possibly associated with poor well integrity, similar to earlier studies (Osborn et al., 2011a). Shallow drinking water could also be con-taminated with deeper saline fluids at the same time as the stray gas migra-tion associated with drilling. A third possibility would be natural migration and connectivity between the shallow drinking water aquifers and deeper, higher salinity formation waters through faults or other more permeable pathways (Warner et al., 2012). This study, in conjunction with a United States Geological Survey (USGS) report using the same major element data (Kresse et al., 2012), are to the authors' knowledge the first to re-port such a comprehensive geochemical evaluation of possible shallow groundwater contamination outside the Marcellus Shale basin (Osborn et al., 2011a and Warner et al., 2012).

5.2 GEOLOGIC SETTING

The study area is located within the currently active development area for the FS in north-central Arkansas with the majority of samples col-lected in Van Buren County and the northern part of Faulkner County (Fig. 1). The area is characterized by a rugged and mountainous landscape to the north and rolling hills to the south, spanning the southern area of the Ozark Mountains, to the northern Arkansas River valley (Imes and Em-mett, 1994). The bedrock in the study area comprises the Pennsylvanian-age Hale, Bloyd and Atoka Formations, which are composed of shale with interbedded minor occurrences of relatively permeable sandstone, lime-stone and coal (Cordova, 1963) (Fig. 2). The shale portion of the Atoka underlies the lowlands because of its lack of resistance to weathering (Cordova, 1963), and thin beds of coal are present throughout but lime-stone is only present in the north of the study area (Cordova, 1963). The Mississippian-age Fayetteville Shale is the target formation during drilling and lies approximately 500-2,100 m below the ground surface (mbgs), with the southern portion of the study area being the deepest. These forma-tions are part of the Western Interior Confining System with groundwater flow restricted to the weathered and fractured upper 100 m of bedrock (Imes and Emmett, 1994). No one formation within this confining sys-tem, even where used for a drinking-water supply, forms a distinct aquifer regionally, and the regional designation as a confining unit indicates that on a regional scale these formations impede the vertical flow of water and confine the underlying aquifers. Domestic wells in the area typically provide limited groundwater yields (Imes and Emmett, 1994). The average reported drinking water well depth is 26 m and minimum and maximum of 7.8 m and 120 m, respectively. Wells drilled deeper than 100 m revealed a much more compacted and less permeable section of the formations (Imes and Emmett, 1994).

The underlying Fayetteville Shale production zone is ~17–180 m thick and occupies an area of approximately 6500 km^2; the area of groundwater samples for this study covered approximately 1/3 of the area of the production zone. The density of shale-gas drilling varied widely across the study area. For the set of drinking-water samples, the total number of unconventional shale-gas wells within 1 km (as measured from the well-head) of a given home ranged from zero to over 14 natural gas wells. This well density represents an area of moderate to intense unconventional shale-gas development similar to other areas of extensive shale gas developments, such as in NE Pennsylvania (Osborn et al., 2011a). Importantly, the Fayetteville Shale is the first oil and gas development in this study area. With no records that indicated the presence of historical conventional wells, which may provide possible conduits for vertical migration of stray gas and/or hydraulic fracturing fluids in other shale-gas plays. Saline water unsuitable for human consumption was identified between 150 and 600 mbgs but generally is at least 300 mbgs in the study area (Imes and Emmett, 1994).

The exposed and shallow subsurface geologic formations serving as local aquifers for Van Buren and Faulkner Counties are a series of dominantly sandstone and shale units of the Hale, Bloyd and Atoka Formations (Fig. 2). Subsurface geology, particularly with respect to lateral facies within the Fayetteville Shale, was poorly defined prior to development of gas, and most of the detailed stratigraphic and reservoir analysis were held as proprietary by the companies operating there.

The Fayetteville Shale is a black, fissile, concretionary shale, which contains pyrite and silica replacement fossils in some intervals. The Fayetteville Shale dips from north to south (Fig. 2). The highly organic-rich facies within the Fayetteville Shale is present in the middle and lower part of the formation. Vitrinite reflectance falls within 1.93–5.09%, which corresponds to the dry gas window (Imes and Emmett, 1994).

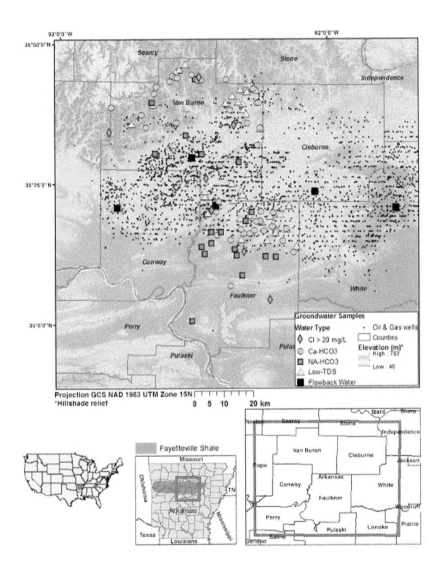

FIGURE 1: Study site location in north-central Arkansas. Unconventional shale-gas wells completed into the Fayetteville Shale are shown in black. Shallow groundwater samples were cataloged based on major element chemistry into four water categories: low-TDS (triangles), Ca–HCO$_3$ (circles), Na–HCO$_3$ (squares), and Cl > 20 mg/L (diamonds).

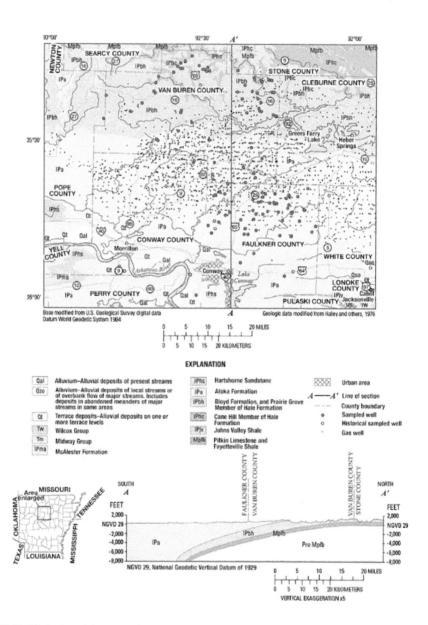

FIGURE 2: (a and b) Map of sample locations and bedrock geology in the study area of north-central Arkansas. The majority of samples were collected from the Atoka (southern area) and Hale Formations (northern area). North-to-south geological cross-section in the study area (A–A′ line is shown). Geological units gently dip to the south with the Atoka Formation outcropping in the southern portion of the study area. The underlying Fayetteville Shale shoals to the north.

The Hale Formation is made up of two members: the lower Cane Hill Member, which is typically composed of silty shale interbedded with siltstone and thin-bedded, fine-grained sandstone, and the upper Prairie Grove Member composed of thin to massive limey sandstone. The Hale Formation thickness is up to 90 m (Imes and Emmett, 1994). The Cane Hill Member of the Hale Formation is exposed in the extreme northern part of Van Buren County (Fig. 2).

The Bloyd Formation in northwestern Arkansas is formally divided into five members, two of which are limestone members absent in the study area. The lower two thirds of the Bloyd Formation consists dominantly of very thin- to thinly-bedded sandstone with shale interbeds. The upper Bloyd is dominantly a shale with interbedded sandstone that is commonly calcareous; the sandstone units can reach a thickness of up to 24 m (Imes and Emmett, 1994). Total thickness for the Bloyd can exceed 120 m in the study area. Exposures of the Bloyd Formation are found in northern Van Buren County (Fig. 2).

The Atoka Formation in the study area consists of a sequence of thick shales that are interbedded with typically thin-bedded, very-fine grained sandstone. The Atoka Formation is unconformable with the underlying Bloyd Formation with a thickness of up to 7500 m in the Ouachita Mountains (Imes and Emmett, 1994). The Atoka Formation is exposed throughout the southern portion of the study area (Fig. 2).

5.3 MATERIALS AND METHODS

All shallow groundwater samples were collected from private drinking water wells by USGS personnel in July and November 2011. Methods for collection of field parameters (pH, temperature, and specific conductance) and water sampling followed standard USGS protocols (Wilde, 2006). These included sampling prior to any holding tanks or filtration, purging water wells until field parameters stabilized, followed by 0.45 µm water filtering on site for water samples collected for trace and major ion analyses. Dissolved gas sample collection followed established protocols (Isotech Laboratories, Inc., 2012). Samples of FS water were collected from production wells (flowback or produced waters) by Arkansas Oil and Gas Commission personnel. Samples were labeled flowback waters if collected within 3 weeks of hydraulic fracturing (5 total samples) and produced water if collected more than 3 weeks after fracturing (1 sample; ~50 weeks following fracturing). All water samples were preserved on ice and shipped to Duke University (Durham, North Carolina, USA), where they were refrigerated until analysis.

Samples for major cations, anions, trace metals, and selected isotopes (O, H, B, Sr and C-DIC) were analyzed at Duke University. Isotech Laboratories performed dissolved gas analysis for concentrations of CH_4 and higher-chain hydrocarbons on 20 samples using chromatographic separation followed by combustion and dual-inlet isotope ratio mass spectrometry to measure $\delta^{13}C_{CH4}$.

Dissolved CH_4 concentrations and $\delta^{13}C\text{-}CH_4$ were determined by cavity ring-down spectroscopy (CRDS) (Busch and Busch, 1997) on an additional 31 samples at the Duke Environmental stable Isotope Laboratory (DEVIL) using a Picarro G2112i. Dissolved CH_4 concentrations were calculated using headspace equilibration, extraction and subsequent concentration calculation by a modification of the method of Kampbell and Vandegrift (1998). For each 1-L sample bottle, 100 mL of headspace was generated by displacing water with zero air (CH_4-free air) injected with gastight syringes equipped with luer-lock valves. Bottles were shaken at 300 rpm for 30 min to equilibrate headspace with dissolved CH_4. The equilibrated headspace was then extracted with gastight syringes while replacing the extracted volume of headspace with deionized water. The extracted headspace was then injected into Tedlar bags (Environmental Supply, Durham, NC) equipped with septum valves and introduced into the Picarro model G2112-i CRDS (Picarro, Inc., Santa Clara, CA). In some cases, dilution into a second Tedlar bag with CH_4-free air (zero-air) was required to keep the measured concentration in the optimal range for the instrument. Calculated detection limits of dissolved CH_4 were 0.002 mg/L water. Reporting limits for reliable $\delta^{13}C\text{-}CH_4$ were 0.1 mg/L, consistent with Isotech Laboratories (Illinois, USA) reporting values. Concentrations and $\delta^{13}C$ values were also corrected for instrument calibrations using known CH4 standards from Airgas (Durham, NC) and Isometric Instruments (Victoria, BC).

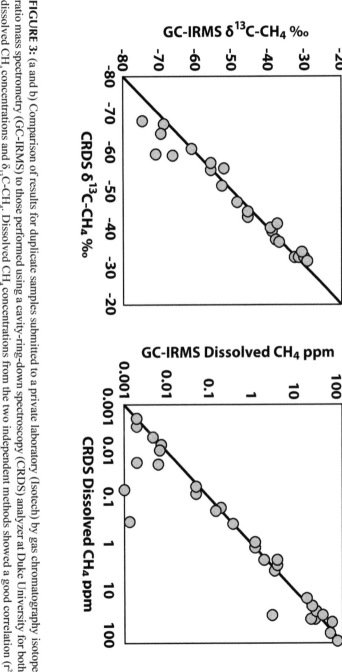

FIGURE 3: (a and b) Comparison of results for duplicate samples submitted to a private laboratory (Isotech) by gas chromatography isotope ratio mass spectrometry (GC-IRMS) to those performed using a cavity-ring-down spectroscopy (CRDS) analyzer at Duke University for both dissolved CH_4 concentrations and $\delta_{13}C$-CH_4. Dissolved CH_4 concentrations from the two independent methods showed a good correlation ($r^2 = 0.90$, $p < 1 \times 10^{-15}$) with some variability at higher concentrations. The comparison of the $\delta^{13}C$-CH_4 values obtained from the two analytical techniques showed a strong correlation ($r^2 = 0.95$, $p < 1 \times 10^{-15}$). The CRDS methodology showed some bias at lower $\delta^{13}C$-CH_4 compared to the private laboratory. Note that this comparison includes samples from other study areas to cover a wide range of concentrations and $\delta^{13}C$-CH_4 values.

To confirm the accuracy of the CRDS results, a set of 49 field duplicate groundwater samples was collected and analyzed at Isotech using gas chromatography isotope ratio mass spectrometer (GC-IRMS). These groundwater samples were collected from North Carolina, New York, Pennsylvania and Arkansas in order to span a wider range of both concentrations (<0.002 mg/L through values well above saturation ~100 mg/L) and C isotope values (−30‰ through −75‰). The comparison of the field duplicates using these two independent methods showed good correlation for concentration (r^2 = 0.90; Fig. 3a) and strong correlation for δ13CCH4 (r^2 = 0.95; Fig. 3b). Relative standard deviation of dissolved CH4 concentrations determined by CRDS on field duplicates was 9.8%. Reproducibility of $\delta^{13}C$ measurements determined by CRDS for 8 field duplicate samples ranged from a minimum of 0.07‰ to a maximum of 1.0‰. Standard deviation of $\delta^{13}C$ measurements (n = 6) on a laboratory check standard was 0.55‰ over the course of the project.

Major anions were determined by ion chromatography, major cations by direct current plasma optical emission spectrometry (DCP-OES), and trace-metals by VG PlasmaQuad-3 inductively coupled plasma mass-spectrometry (ICP-MS). Four replicate samples showed good reproducibility (<5%) for both major and trace element concentrations. Strontium and B isotopes were determined by thermal ionization mass spectrometry (TIMS) on a ThermoFisher Triton at the TIMS laboratory in Duke University. The average $^{87}Sr/^{86}Sr$ of the SRM-987 standard measured during this study was 0.710266 ± 0.000005 (SD). The average $^{11}B/^{10}B$ of NIST SRM-951 during this study was 4.0055 ± 0.0015. The long-term standard deviation of $\delta^{11}B$ in the standard and seawater replicate measurements was 0.5‰. DIC concentrations were determined in duplicate by titration with HCl to pH 4.5. Values of $\delta^{18}O$ and δ^2H of water were determined by thermochemical elemental analysis/continuous flow isotope ratio mass spectrometry (TCEA-CFIRMS), using a ThermoFinnigan TCEA and Delta + XL mass spectrometer at DEVIL. $\delta^{18}O$ and δ^2H values were normalized to V-SMOW and V-SLAP. The C isotope ratio of dissolved inorganic C detemined after acid digestion, on a ThermoFinnigan (Bremen, Germany) GasBench II feeding a ThermoFinnigan Delta + XL Isotope Ratio Mass Spectrometer (IRMS) in the DEVIL lab. Several mL (volume depending on

DIC concentration) of each sample were injected into 11-mL septum vials that had each been pre-dosed with 150 uL phosphoric acid and pre-flushed for 10 min with He at 50 mL/min to remove air background. Raw $\delta^{13}C$ of resulting CO_2 was normalized vs Vienna Pee Dee Belemnite (VPDB) using NBS19, IAEA CO-8 standards, and an internal $CaCO_3$ standard.

Natural gas well locations (representing locations of the vertical portion of the well) were obtained from the Arkansas Oil and Gas Commission database (Arkansas Oil and Gas Commission, 2012). Arkansas Oil and Gas Commission also provided ^{228}Ra and ^{226}Ra values for five flowback and one produced water sample. Historical water data were gathered from the USGS National Water Information System (NWIS) data base for the six counties that comprise the bulk of permitted and active gas production wells: Cleburne, Conway, Faulkner, Independence, Van Buren and White Counties (Fig. 2). The data set includes 43 groundwater samples collected near the study area prior to shale-gas development during 1948 and 1983 (USGS, 2013).

5.4 RESULTS AND DISCUSSION

5.4.1 GEOCHEMICAL CHARACTERIZATION OF THE SHALLOW GROUNDWATER

The 127 shallow groundwater samples were divided into four major water categories (Fig. 1 and Supplementary data). The first category was low-TDS (<100 mg/L) and generally low-pH (pH < 6.6; n = 54) water. The second was a Ca–HCO_3 water (n = 40), with moderate TDS (100 > TDS < 200 mg/L). The third was a Na–HCO_3 water with a wider range of TDS (100 > TDS < 415 mg/L; n = 24). The fourth group was classified as Ca–Na–HCO_3 water type with the highest TDS (200 > TDS < 487 mg/L) and slightly elevated Cl (>20 mg/L) and Br/Cl molar ratios >1 × 10^{-3} (n = 9). The fourth group was identified because the elevated Cl and Br/Cl could potentially indicate contamination from the underlying saline formation water (see description below).

The C isotope ratio of dissolved inorganic C ($\delta^{13}C_{DIC}$; n = 81 samples) ranged from −22‰ to −10‰ (Supplementary data). The low-TDS and Ca–HCO$_3$ water types had lower DIC concentrations but all water types had similar $\delta^{13}C_{DIC}$, while most water samples fell within a narrower and lower range of −20‰ to −17‰. In the Na–HCO$_3$ groundwater a positive correlation was observed between DIC concentrations and $\delta^{13}C_{DIC}$ values (r^2 = 0.49, p < 0.05; Fig. 4). The Sr isotope ratios ($^{87}Sr/^{86}Sr$) varied from 0.7097 to 0.7166 (Fig. 5a). Most of the Ca–HCO$_3$ waters had slightly lower $^{87}Sr/^{86}Sr$ (mean = 0.71259; n = 12) relative to the Na–HCO$_3$ waters (mean = 0.71543; n = 13). Boron isotope ratios ($\delta^{11}B$) showed a wide range from 4‰ to 33‰, with a general increase of $\delta^{11}B$ with B content (Fig. 5b) with no systematic distinction between the water types (p > 0.05). The stable isotope composition of all water types did not show any distinctions (p > 0.05) related to the water composition (Supplementary data) and $\delta^{18}O$ and δ^2H variations are consistent with the Local Meteoric Water Line (LMWL) (Kendall and Coplen, 2001) of modern precipitation in the region. This similarity suggests a common meteoric origin, and also indicates that all of the geochemical modifications presented below were induced from water–rock interactions along groundwater flowpaths in the shallow aquifers.

Historical groundwater quality data from in or near the study area from the NWIS data base (Fig. 2) included 43 samples collected prior to shale-gas development between 1948 and 1983 (Table 1). Although collected from the same formations, the majority of historical samples were collected to the east, and only three sampling sites overlapped with the intensely sampled part of the study area (Fig. 2); therefore, a complete statistical comparison to historical data was not possible. However, the reported chemical composition of the water samples collected prior to shale gas development in the area was consistent with the Ca–HCO$_3$ and Na–HCO$_3$ water types, with a predominance of Na–HCO$_3$ water type in the Atoka Formation (Fig. 2) as reported in previous studies (Cordova, 1963). Likewise, the range of concentrations in this study fell within the minimum and maximum reported values in the NWIS (Table 1).

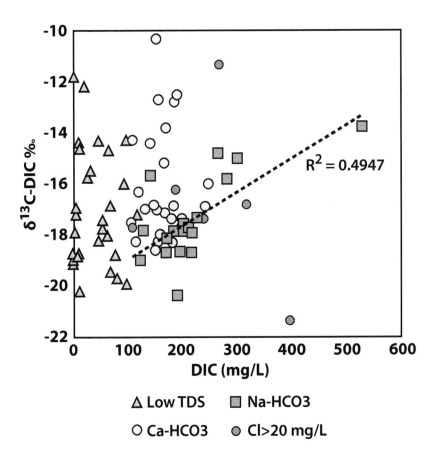

FIGURE 4: δ^{13}C-DIC (‰) and DIC (mg/L) in shallow groundwater samples. The average δ^{13}C-DIC (−17‰ to −20‰) in the bulk groundwater indicates the majority of DIC is derived from weathering of silicate minerals that would approach −22‰. Methanogens in some of the Na–HCO$_3$ waters would leave DIC with elevated residual δ^{13}C-DIC (line).

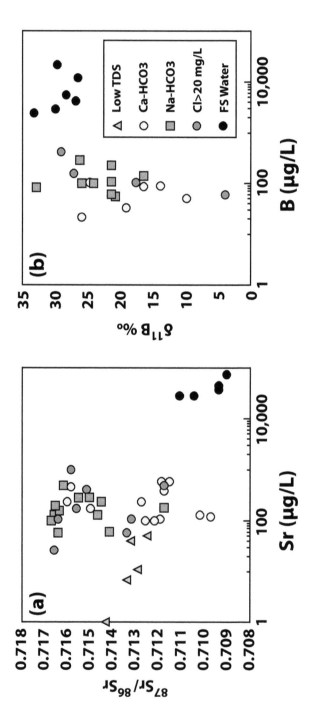

FIGURE 5: (a and b) $^{87}Sr/^{86}Sr$ versus Sr concentration (μg/L) log scale and $\delta^{11}B‰$ versus B concentration (μg/L) in log scale. The lack of strong Sr and B isotopic relationships exclude possible mixing between the Fayetteville Shale water and the shallow groundwater. Instead the isotopic variations appear to be controlled by weathering and water–rock interaction.

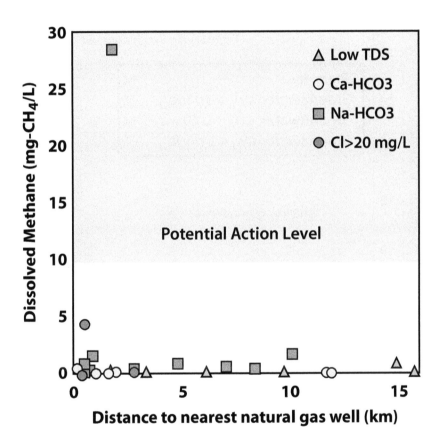

FIGURE 6: Dissolved CH$_4$ concentrations (mg/L) in domestic wells plotted versus distance from the domestic wells to nearest natural gas well. Only one of 51 wells analyzed contained CH$_4$ at concentrations above the potential action level set by the Department of Interior (10 mg/L). There is no statistically significant difference in dissolved CH$_4$ concentrations from wells collected within 1 km of a gas well and those collected >1 km from a well. The highest dissolved CH$_4$ concentrations were detected in Na–HCO$_3$ water.

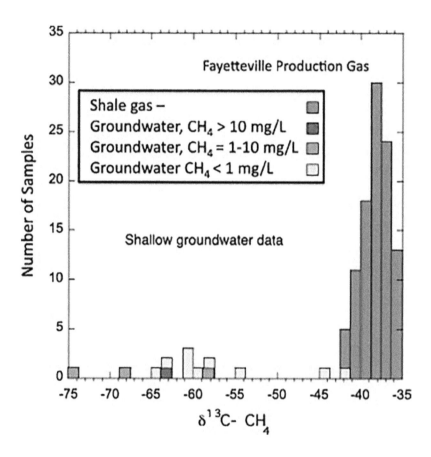

Figure 7: Histogram of $\delta^{13}C\text{-}CH_4$ (‰) values of dissolved CH_4 plotted in comparison to published values for Fayetteville Shale produced gas $\delta^{13}C\text{-}CH_4$ (‰) (Zumberge et al., 2012). Concentrations of the dissolved CH_4 in the studiedshallow groundwater samples are indicated by color. The majority of samples, including all of those at higher CH_4 concentrations plot at more negative $\delta^{13}C\text{-}CH_4$ values, indicating that a shallow biogenic origin likely contributes to the formation of CH_4. The lone sample that overlaps with Fayetteville Shale values may represent migration of stray production gas, but at very low concentrations.

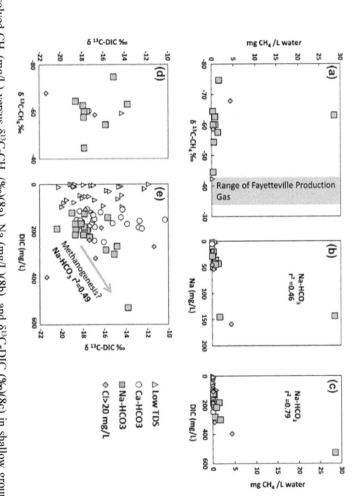

Figure 8: Dissolved CH$_4$ (mg/L) versus δ^{13}C-CH$_4$ (‰)(8a), Na (mg/L)(8b), and δ^{13}C-DIC (‰)(8c) in shallow groundwater samples. The correlations observed between CH$_4$ and Na (r^2 = 0.46) and DIC (r^2 = 0.79) indicate that the highest CH$_4$ is found in Na–HCO$_3$ groundwater. At the higher DIC and CH$_4$ concentrations the depleted δ^{13}C-CH$_4$ indicates that methanogens likely contribute to the formation of CH$_4$. δ^{13}C-DIC versus δ^{13}C-CH$_4$ (‰)(8d) and DIC (mg/L)(8e) in shallow groundwater. The average δ^{13}C-DIC (−17‰ to −20‰) in the bulk groundwater indicates the majority of DIC is derived from weathering of silicate minerals that would approach −22‰. Methanogens in some of the Na–HCO$_3$ waters would leave DIC with elevated residual δ13C-DIC (arrow).

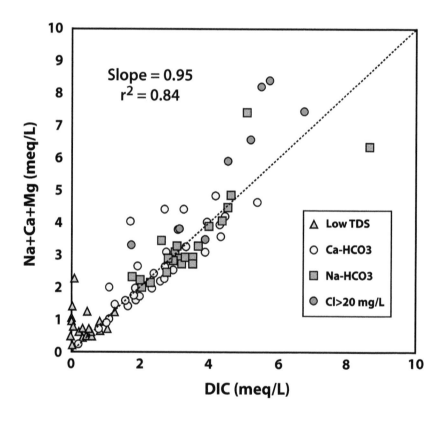

FIGURE 9: The sum of Na, Ca and Mg (meq/L) versus dissolved inorganic C (DIC; meq/L) in shallow groundwater samples. Note that DIC balances the majority of the total cations in shallow groundwater samples across all water types.

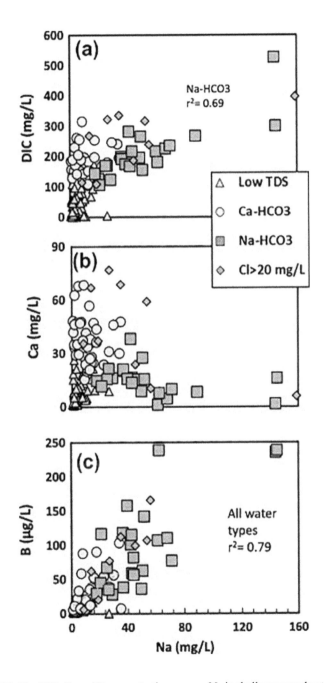

FIGURE 10: The DIC, Ca and B concentrations versus Na in shallow groundwater samples.

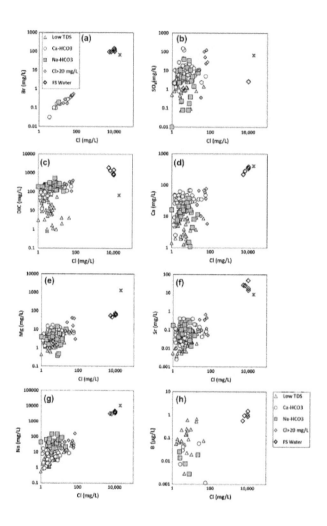

FIGURE 11: The variations of major elements as normalized to Cl⁻ in shallow groundwater and the FS saline water. The composition of the FS water infers modified seawater through evaporation and halite precipitation (high Br/Cl ratio), water–rock interactions (enrichment of Na, Sr, Mg and Ca relative to the expected evaporated seawater curve), followed by dilution with meteoric water. Note that there is no apparent relationship between concentrations of constituents in shallow groundwater and the deeper FS waters. The negative correlation between Cl and DIC indicates that dilution is the main factor for the high DIC in the formation water. The positive correlation of Na/Cl and Ca/Cl with DIC concentration indicates that Na, Ca and DIC within the FS are likely sourced from carbonate dissolution combined with base-exchange reactions that have modified the original composition of the FS water.

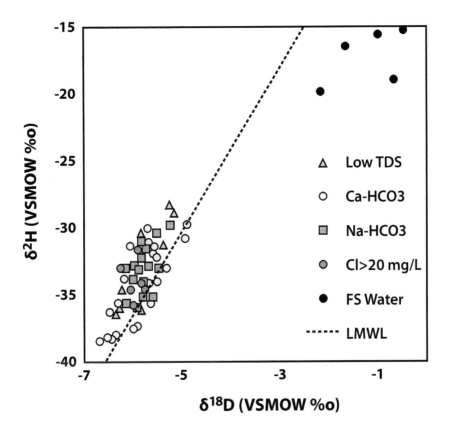

FIGURE 12: δ^2H versus δ^{18}O values in shallow groundwater and the Fayetteville Shale brines. The relationship between δ^{18}O and δ^2H in shallow groundwater is consistent with the local meteoric water line (LMWL) while the Fayetteville Shale brines plot to the right of the LMWL and could reflect mixing between depleted δ^{18}O and δ^2H low-saline water and δ^{18}O and δ^2H-enriched brines.

5.4.2 METHANE SOURCES IN SHALLOW GROUNDWATER

Dissolved CH_4 concentrations were determined in 51 of the 127 water samples from wells collected for this study (Supplementary data). Methane was detected (>0.002 mg/L) in 63% of wells (32 of the 51), but only six wells had concentrations >0.5 mg CH_4/L, with a single sample point (28.5 mg/L) above the potential recommended action level in the USA [10 mg/L] (Eltschlager et al., 2001) (Fig. 6). Dissolved CH_4 concentrations were not higher closer to shale gas wells (Fig. 6 and Supplementary data), nor was any statistical difference (student t-test) apparent between concentrations in groundwater of 32 wells collected within 1 km of shale-gas production and 19 wells >1 km away from gas wells (p > 0.1; Supplementary data).

The C isotope ratios of CH4 ($\delta^{13}C_{CH4}$) was measurable in 14 of 51 samples (dissolved CH_4 > 0.1 mg/L) and ranged from −42.3‰ to −74.7‰ (Fig. 7), but the range in $\delta^{13}C_{CH4}$ in the six samples with concentrations greater than 0.5 mg/L was systematically (p < 0.01) lower (−57.6‰ to −74.7‰). This result provides evidence for a predominantly biogenic origin of the dissolved gas (i.e., <−55‰) (Coleman et al., 1981, Whiticar et al., 1986, Grossman et al., 1989 and Whiticar, 1999). Additionally, $\delta^{13}C_{CH4}$ of 13 out of 14 samples with measurable $\delta^{13}C_{CH4}$ did not overlap the reported values (Zumberge et al., 2012) for Fayetteville Shale production gas (Fig. 7). The only one sample with a $\delta^{13}C_{CH4}$ value (−42.3‰) that approaches the values reported for shale gas had low CH_4 concentration (0.15 mg/L).

Samples with trace (<0.5 mg/L) CH_4 concentrations and $\delta^{13}C_{CH4}$ values between −42 and −60‰ could reflect either flux of deep-source thermogenic gas (Schoell, 1980) or a mixture of biogenic and thermogenic gas. The sample with the highest $\delta^{13}CC_{H4}$ value (−42.3‰) also had a low Cl^- concentration (2 mg/L). The combined low Cl^- and CH_4 concentrations rule out likely contamination from underlying fluids (gas and water) (see discussion below). Further evidence for a biogenic origin of CH_4 in the shallow groundwater was provided by the lack of detectable higher chain hydrocarbons (C_2 < 0.0005 mol%) in the 20 samples analyzed at the commercial laboratory (C_2+ was detected in only 1 of 20 samples analyzed). The single detection of a higher chain hydrocarbon (C_2 = 0.0277 mol%) was in a sample with a relatively high C_1/C_2 + ratio = $(C_1/C_2$ =

730), consistent with a biogenic source (~1000) (Schoell, 1980 and Coleman et al., 1981) (Supplementary data). The distribution of dissolved CH_4 concentrations and $\delta^{13}C_{CH4}$ values (Fig. 8a) suggest a local, shallow origin of dissolved CH_4 unrelated to shale-gas extraction in the vast majority of samples.

If the CH_4 was sourced from biogenic processes within the shallow aquifers, the ground water chemistry should provide further support for its biogenic origin (Aravena et al., 1995). Median dissolved CH_4 concentrations were highest in the Na–HCO_3 water type, with positive correlations to Na and DIC ($r^2 = 0.46$ and 0.79, respectively; Fig. 8b and c). In addition, the positive correlation between DIC concentrations and $\delta^{13}C_{DIC}$ values ($r^2 = 0.49$, $p < 0.05$; Fig. 8e) could suggest that methanogenesis is occurring within the formations, perhaps within the minor coal beds (Imes and Emmett, 1994) under reducing conditions. If the minor concentrations of observed CH_4 were sourced from microbial CO_2 reduction, generation would be expected of $\delta^{13}C_{CH4}$ of $-70‰$ to $-80‰$ (Coleman et al., 1981) parallel to elevated residual $\delta^{13}C_{DIC}$ (e.g., $>+10‰$) during CH_4 production (Aravena and Wassenaar, 1993). However, in the study the majority of the $\delta^{13}C_{DIC}$ values are significantly lower ($-20‰$ to $-17‰$), demonstrating that methanogens are not the main control of $\delta^{13}C_{DIC}$ in the aquifer. In the low TDS water only trace levels ($CH_4 < 0.8$ mg/L) of dissolved CH_4 were recorded ($n = 9$) and only two low-TDS samples had detectable higher $\delta^{13}C_{CH4}$ (-42.3 and $-59.6‰$). These values could indicate either a minor presence of thermogenic gas in the shallow aquifers or bacterial oxidation (Coleman et al., 1981).

5.4.3 WATER–ROCK INTERACTIONS AND MIXING WITH EXTERNAL FLUIDS

The geochemical variation from low TDS, Ca–HCO_3 and Na–HCO_3 water types infer different modes of water–rock interaction. The low-TDS waters could reflect an early stage of groundwater recharge without much mineralization induced from water–rock interaction, while the Ca–HCO_3 waters suggest dissolution of carbonate minerals in the aquifers. A Na–HCO_3 water type typically (e.g., Cheung et al., 2010) indicates silicate weathering and ion exchange processes (e.g., reverse base-exchange reaction). In the majority of the shallow groundwater samples, regardless of the water type, DIC nearly balanced the sum of Na, Ca and Mg concentrations (in equivalent units; Fig. 9). DIC could be generated in the shallow aquifers by weathering of silicate minerals in the shale, dissolution of marine carbonate by H_2CO_3 produced through oxidation of organic matter, or bacterial SO_4 reduction. Silicate weathering would mobilize Na, Ca, Mg and Sr with a radiogenic $^{87}Sr/^{86}Sr$ signature and B with a wide $\delta^{11}B$ range from 0‰, which would characterize structural B in silicate minerals (Lemarchand and Gaillardet, 2006), to 15–20‰ in "desorbable" B on marine clay surfaces (Spivack and Edmond, 1987). The $\delta^{13}C_{DIC}$ value would reflect the isotopic fractionation between DIC species and would be expected to be similar to the composition of the H_2CO_3 that triggered the silicate weathering ($\sim -22‰$). If, instead, dissolution of marine carbonate minerals was occurring, one would expect contributions of Ca, Mg and Sr with a low $^{87}Sr/^{86}Sr$ (~ 0.7082) for the Pennsylvanian-age marine formation (Burke et al., 1982), and B with $\delta^{11}B$ of a marine carbonate signature ($\sim 20‰$) (Vengosh et al., 1991). Dissolution of marine carbonate would generate View the MathML source with $\delta^{13}C_{DIC} \sim -11‰$, assuming a closed system with equal proportions of marine calcite dissolution ($\delta^{13}C_{DIC} \sim 0‰$) and H_2CO_3 ($\delta^{13}C_{DIC} \sim -22‰$), and that all DIC-bearing species would be in isotopic equilibrium (McCaffrey et al., 1987). Carbonate dissolution could contribute Ca that would be exchanged with Na from exchange sites on clay minerals, resulting in Na–HCO_3 water. In such a scenario, the Ca concentrations would be inversely correlated with Na.

　　Examining all of these geochemical and isotopic constraints, it is clearly shown that neither of these two mechanisms (i.e., silicate weathering versus marine carbonate dissolution combined with base-exchange reaction) were explicitly consistent with the geochemical variations measured in the shallow groundwater in this study. For example, in most of the groundwater samples, including those defined as the Na–HCO_3 type, Na was positively correlated with Ca, indicating contributions from both elements that would reflect silicate weathering. In contrast, the most DIC-rich (Fig. 10a) waters showed an inverse relationship between Na and Ca (Fig. 10b) that typically mimics reverse base-exchange reactions. Likewise, all of the water types showed a positive correlation ($r^2 = 0.79$) between Na and B concentrations (Fig. 10c), a combination that could reflect mobilization from exchangeable sites on clay minerals. The most DIC-rich waters have a lower Ca/Na ratio and lower Na relative to B concentration (Fig. 10c), inferring a different source. The $\delta^{11}B$ of the Na–HCO_3 waters (16.5–33‰; Fig. 5b) was also consistent with B sourced from exchangeable sites on marine clay minerals.

In contrast, relatively low $\delta^{13}C_{DIC}$ (−20‰ to −17‰) (Supplementary data) and radiogenic $^{87}Sr/^{86}Sr$ ratios (0.7097–0.7166) (Fig. 5a) in the majority of the studied groundwater rule out the possibility that marine carbonate dissolution was the major process that controlled the generation of Ca–HCO_3 water. Nonetheless, given that the shale in the study area is carbonate-rich (Imes and Emmett, 1994), carbonate dissolution likely contributed Ca and HCO_3, with Ca exchanged with Na to generate Na–HCO_3 water. Reverse base-exchange reaction would remove Ca and Sr, and the uptake of Sr is not expected to modify its original isotopic ratio (i.e., $^{87}Sr/^{86}Sr$ ratio of the Pennsylvanian-age marine carbonate). One possible explanation for the high $^{87}Sr/^{86}Sr$ ratio is that the carbonate in the shale was diagenetically-modified from bacterial SO_4 reduction with modified fluids containing radiogenic $^{87}Sr/^{86}Sr$ and depleted $\delta^{13}C_{DIC}$ relative to the original composition of the marine carbonates. Given that the groundwater has a radiogenic $^{87}Sr/^{86}Sr$ ratio (0.7097–0.7166) that is similar to the composition of the local shale formations (Kresse and Hays, 2009), it was concluded that the water chemistry was controlled by both silicate mineral weathering and dissolution by diagenetically modified carbonate cement followed by ion-exchange reactions. Further study is needed to characterize the composition of the carbonate cement and delineate the specific mechanism that has caused evolution of the groundwater into a Na-HCO_3 composition.

The fourth shallow groundwater type, the higher-Cl^- waters, shows a strong correlation between Cl and Br ($r^2 = 0.89$; Fig. 11a) with a high Br/Cl ratio ($>1 \times 10^{-3}$) that is similar to the elevated Br/Cl in the FS brine (see below). This geochemical composition could be interpreted as mixing of shallow groundwater with underlying formation water, similar to the salinization phenomena observed in NE Pennsylvania (Warner et al., 2012). However, the variations of other dissolved constituents such as B and Sr were not correlated with Cl^- (Fig. 11f–h), and their isotopic ratios, including $^{87}Sr/^{86}Sr$ (Fig. 5a), $\delta^{13}C_{DIC}$ (Fig. 4), and the majority of $\delta^{11}B$ (Fig. 5b) were distinctly different from expected mixing relations with the FS brines (Supplementary data). This infers that the composition of the groundwater with (Cl > 20 mg/L) was modified by weathering and water–rock interaction. The ability to delineate the exact saline end-member that generated the saline groundwater is limited.

Finally, neither the Na–HCO$_3$ water type, nor the fourth water type with Cl > 20 mg/L were located closer to shale-gas wells (Supplementary data), which rules out the likelihood of salinization induced from shale gas exploitation and migration of fluids associated with natural gas wells. Instead, a geographical distribution of the water types was observed; the majority of Na–HCO$_3$ samples were identified in the southern portion of the study area (Fig. 1) and at lower average elevations (Supplementary data), which could indicate increased Na and DIC in the southern portion of the study area, corresponding to a regional groundwater flow and increased water–rock interaction along regional flow paths (Imes and Emmett, 1994 and Kresse et al., 2012) and/or greater predominance of shale lithology in the low lying regions (Cordova, 1963).

5.4.4 THE FAYETTEVILLE SHALE FLOWBACK AND PRODUCED WATERS

The FS flowback and produced water samples (Supplementary data) were saline (TDS ~20,000 mg/L), yet the present data show that the salinity is substantially lower than produced waters from other shale gas basins (e.g., Marcellus brine with TDS ~200,000 mg/L; Table 2). The FS saline water was composed of Na–Cl–HCO$_3$, with a linear correlation ($r^2 = 0.39$) between Cl$^-$ and Br$^-$ and a high Br/Cl ratio (~4 × 10^{-3} to 7 × 10^{-3}; Fig. 11a). This composition infers modified evaporated seawater (seawater evaporation, salt precipitation, followed by dilution with meteoric water) with Na, Sr, Mg and Ca enrichments relative to the expected evaporated seawater curve (McCaffrey et al., 1987) (Fig. 11a–h). The $\delta^{18}O$ (−2.1‰ to −0.5‰) and δ^2H (−19.8‰ to −15.2‰) of the formation water samples plot to the right of the $\delta^2H/\delta^{18}O$ LMWL (Kendall and Coplen, 2001) (Fig. 12). DIC content was elevated (800–1800 mg/L) compared to other produced waters in other shale basins in the USA (Table 2), and had a distinctive elevated $\delta^{13}C_{DIC}$ (−12.7‰ to +3.7‰), which may reflect the composition of the injected hydraulic fracturing fluid or methanogenesis. Boron ($\delta^{11}B$ = 26–30‰; Supplementary data and Fig. 5b) and Sr ($^{87}Sr/^{86}Sr$ = 0.7090–0.7111; Supplementary data and Fig. 5a) isotopic fingerprints were different than would be expected for unaltered Mississippian-age evaporated seawater, which would generate $\delta^{11}B > 39‰$ and a less radiogenic $^{87}Sr/^{86}Sr$ ratio of ~0.7082 (Burke et al., 1982). The ^{226}Ra and ^{228}Ra activities were relatively low (14-260 pCi/L) (Supplementary data) compared to Appalachian brines (Rowan et al., 2011) with a $^{228}Ra/^{226}Ra$ range of 0.1–0.5, which is similar to Appalachian brines. This relatively low Ra level could have important implications for management strategies and evaluation of possible environmental effects, following disposal of the flowback and produced waters.

TABLE 2: Typical produced water TDS (mg/L) concentrations.

	TDS (mg/L)	DIC (mg/L)
Fayetteville Shale	25,000	1300[a]
Barnett Shale	60,000	610[b]
Woodford Shale	110,000	
Haynesville Shale	120,000	
Permian basin	140,000	
Marcellus Shale	180,000	140[b]

[a]*This study. b EPA workshop on hydraulic fracturing – http://www.epa.gov/hfstudy/12_ Hayes_-_Marcellus_Flowback_Reuse_508.pdf. Source: Kimball, 2012 citation of USGS produced water database – available at http://energy.cr.usgs.gov/prov/prodwat/data.htm.*

The chemical composition of the five flowback samples reflects mixing between the original formation water (represented by the produced water) and lower-saline water that was injected as fracturing fluids. Given the higher salinity of the formation water (relative to the injected water) its chemistry overwhelmingly controlled the composition of the flowback waters. Similar results were observed in the composition of flowback water from the Marcellus Formation (Haluszczak et al., 2013). Overall, the combined geochemical data from five flowback and one produced water samples indicate that the FS water is likely the remnant of seawater that evaporated beyond the halite saturation stage (McCaffrey et al., 1987). Similar to the Appalachian brines (Dresel and Rose, 2010 and Warner et al., 2012) the evaporated seawater was modified by water–rock interaction that resulted in Na, Sr, Mg and Ca enrichments and alterations of the original marine $\delta^{11}B$ and $^{87}Sr/^{86}Sr$ isotopic fingerprints. The brine was subsequently diluted by meteoric water with lower $\delta^{18}O$ and δ^2H values that reduced the original salinity to levels lower than seawater (TDS < 32,000 mg/L).

Another unique characteristic of the FS is the substantial DIC enrichment that is inversely correlated (r^2 = 0.55) with Cl^- content (Fig. 11c) with high $\delta^{13}C_{DIC}$ values ($-12.7‰$ to $3.7‰$) relative to shallow groundwater (Supplementary data). This suggests that the FS water is diluted with DIC-rich water. The elevated positive $\delta^{13}C_{DIC}$ could infer methanogenesis in the low-saline water that diluted the original FS brine. Alternatively, dissolution of the limestone matrix with a $\delta^{13}C$ of ~1.0‰ (Handford, 1986) coupled with reverse base-exchange reaction within the FS would generate Ca, Na (from base-exchange) and DIC with a positive $\delta^{13}C_{DIC}$ signature. This is confirmed by the correlation of Na/Cl and Ca/Cl ratios and inverse correlation of Cl with DIC (Fig. 11g, d and c). Combined, the chemistry and isotopic results indicate a major modification and dilution of the original FS brine composition.

5.5 CONCLUSIONS AND IMPLICATIONS

This study examined water quality and hydrogeochemistry in groundwater from shallow aquifers in an attempt to delineate possible groundwater contamination. Three types of potential contamination were considered (1) stray gas contamination; (2) migration of saline fluids from depth that were directly associated with drilling and exploration of the underlying Fayetteville Shale; and (3) natural migration of saline fluids from depth through permeable geological formations. The results of this study clearly show lack of saline fluid contamination (scenario #2) in drinking water wells located near shale gas sites, which is consistent with previous studies in shallow groundwater in the Marcellus in northeastern Pennsylvania (Osborn et al., 2011a and Warner et al., 2012). However, the lack of apparent CH_4 contamination with thermogenic C isotope composition in shallow groundwater near shale gas sites in the Fayetteville Shale differs from results reported for shallow groundwater aquifers overlying the Marcellus Formation (Osborn et al., 2011a). It has been proposed that the stray gas contamination likely resulted from poor well integrity that allowed leakage and migration of CH_4 to the shallow aquifers (Jackson et al., 2011, Osborn et al., 2011a and Osborn et al., 2011b). In this study no direct evidence was found for stray gas contamination in groundwater wells located near shale gas sites and most of the CH_4 identified (mostly low concentrations) had a $\delta^{13}C_{CH4}$ composition that is different from the fingerprint of the Fayetteville Shale gas.

Likewise, this study did not find geochemical evidence for natural hydraulic connectivity between deeper formations and shallow aquifers (Warner et al., 2012) that might provide conduits for flow of saline fluids from depth to the shallow groundwater. The spatial distribution of the slightly saline groundwater (Cl > 20 mg/L) that could be derived from dilution of the FS brine or another saline source was not associated with the location of the shale gas wells. Shallow groundwater samples for this study were collected from formations that are part of the Western Interior Confining System (Imes and Emmett, 1994). A previous investigation has shown that these formations impede the vertical flow of groundwater and restrict groundwater movement for domestic supply wells to only local near-surface flow systems (Imes and Emmett, 1994). The natural impermeability and apparent lack of deformation of these formations seems to prevent hydraulic connectivity that might allow the flow of saline fluids between deep saline formations and shallow drinking water aquifers in north-central Arkansas.

The lack of fracture systems that would enable hydraulic connectivity is very different from the geological formations overlying the Marcellus Shale in the Appalachian basin (Warner et al., 2012 and references therein). These differences could be explained by two structural deformation scenarios: (1) recent glaciation and isostatic rebound of shallow bedrock that was reported in the Appalachian and Michigan basins (Weaver et al., 1995); and (2) tectonic deformation that shaped particularly the Appalachian Basin (Lash and Engelder, 2009). These natural deformation events could explain the increased hydraulic connectivity and pathways that provide conduits for fluids and gas between the deeper production zones and shallow groundwater in the shallow geological formations overlying the Marcellus Shale in the Northern Appalachian Basin but apparently not in the study area in Arkansas.

Previous studies in the Marcellus Basin have suggested that the CH_4 leakage to shallow drinking water wells is most likely attributable to poor well integrity (Osborn et al., 2011a). Such human factors could also explain the lack of CH_4 contamination in Arkansas, possibly due to: (1) better wellbore integrity; and/or (2) a lack of conventional oil and gas development in north-central Arkansas prior to the shale gas extraction from the Fayetteville Formation (Kresse et al., 2012).

In conclusion, this study demonstrates the importance of basin- and site-specific investigations in an attempt to determine the possible effects of shale gas drilling and hydraulic fracturing on the quality of water resources. The study shows that possible groundwater impacts from shale-gas development differ between basins and variations in both local and regional geology could play major roles on hydraulic connectivity and subsurface contamination processes. Based on the results of this and previous studies (Osborn et al., 2011a and Warner et al., 2012), it is concluded that systematic monitoring of multiple geochemical and isotopic tracers is necessary for assessing possible groundwater contamination in areas associated with shale gas exploration as well as the possible hydraulic connectivity between shallow aquifers and deeper production zones.

REFERENCES

1. R. Aravena, L.I. Wassenaar. Dissolved organic carbon and methane in a regional confined aquifer, southern Ontario, Canada: carbon isotope evidence for associated subsurface sources. Appl. Geochem., 8 (1993), pp. 483–493
2. R. Aravena, L.I. Wassenaar, J.F. Barker. Distribution and isotopic characterization of methane in a confined aquifer in southern Ontario, Canada. J. Hydrol., 173 (1995), pp. 51–70
3. Arkansas Oil and Gas Commission, 2012. Arkansas Online Data System. <http://www.aogc.state.ar.us/JDesignerPro/JDPArkansas/AR_Welcome.html>.

4. W.H. Burke, R.E. Denison, E.A. Hetherington, R.B. Koepnick, H.F. Nelson, J.B. Otto. Variation of seawater 87Sr/86Sr throughout Phanerozoic time. Geology, 10 (1982), pp. 516–519

5. Busch, K., Busch, M., 1997. Cavity ring-down spectroscopy: An ultratrace absorption measurement technique. Am. Chem. Soc. Symp. Series 720, Oxford.

6. K. Cheung, P. Klassen, B. Mayer, F. Goodarzi, R. Aravena. Major ion and isotope geochemistry of fluids and gases from coalbed methane and shallow groundwater wells in Alberta, Canada. Appl. Geochem., 25 (2010), pp. 1307–1329

7. D.D. Coleman, J.B. Risatti, M. Schoell. Fractionation of carbon and hydrogen isotopes by methane-oxidizing bacteria. Geochim. Cosmochim. Acta, 45 (1981), pp. 1033–1037

8. R. Cordova. Water Resources of the Arkansas Valley Region, Arkansas. US Geological Survey, Washington, DC (1963)

9. Dresel, P., Rose, A., 2010. Chemistry and origin of oil and gas well brines in western Pennsylvania. Pennsylvania Geol. Surv., 4th series Open-File Report OFOG 10-01.0. Pennsylvania Department of Conservation and Natural Resources.

10. Eltschlager, K., Hawkins, J., Ehler, W., Baldassare, F., 2001. Technical measures for the investigation and mitigation of fugitive methane hazards in areas of coal mining. U.S. Department of the Interior, Office of Surface Mining Reclamation and Enforcement.

11. E.L. Grossman, B.K. Coffman, S.J. Fritz, H. Wada. Bacterial production of methane and its influence on ground-water chemistry in east-central Texas aquifers. Geology, 17 (1989), pp. 495–499

12. L.O. Haluszczak, A.W. Rose, L.R. Kump. Geochemical evaluation of flowback brine from Marcellus gas wells in Pennsylvania, USA. Appl. Geochem., 28 (2013), pp. 55–61

13. C. Handford. Facies and bedding sequences in shelf-storm deposited carbonates—Fayetteville Shale and Pitkin Limestone (Mississippian), Arkansas. J. Sed. Petrol., 56 (1986), pp. 123–137

14. Imes, J., Emmett, L., 1994. Geohydrology of the Ozark Plateaus aquifer system in parts of Missouri, Arkansas, Oklahoma, and Kansas. U.S. Geol. Surv. Prof. Paper 1414D.

15. R.B. Jackson, S.G. Osborn, A. Vengosh, N.R. Warner. Reply to Davies: hydraulic fracturing remains a possible mechanism for observed methane contamination of drinking water. Proc. Nat. Acad. Sci., 108 (2011), p. E872

16. D.H. Kampbell, S.A. Vandegrift. Analysis of dissolved methane, ethane, and ethylene in ground water by a standard gas chromatographic technique. J. Chromatog. Sci., 36 (1998), pp. 253–256

17. D.M. Kargbo, R.G. Wilhelm, D.J. Campbell. Natural gas plays in the marcellus shale: challenges and potential opportunities. Environ. Sci. Technol., 44 (2010), pp. 5679–5684

18. C. Kendall, T. Coplen. Distribution of oxygen-18 and deuterium in river waters across the United States. Hydrol. Process., 15 (2001), pp. 1363–1393

19. R.A. Kerr. Natural gas from shale bursts onto the scene. Science, 328 (2010), pp. 1624–1626

20. Kimball, R., 2012. Key Considerations for Frac Flowback/Produced Water Reuse and Treatment. NJ Water Environment Association Ann. Conf. Atlantic City, NJ, May 2012. <http://www.aaees.org/downloadcenter/2012NJWEAPresentation-RobertKimball.pdf>.

21. Kresse, T., Hays, P., 2009. Geochemistry, comparative analysis, and physical and chemical characteristics of the thermal waters east of Hot Springs National Park, Arkansas, 2006–09. U.S. Geol. Surv. Scient. Invest. Rep. 2009–5263.

22. Kresse, T., Warner, N., Hays, P., Down, A., Vengosh, A., Jackson, R., 2012. Shallow groundwater quality and geochemistry in the Fayetteville Shale gas-production area, north-central Arkansas, 2011. U.S. Geol. Surv. Scient. Invest. Rep. 2012–5273.

23. G.G. Lash, T. Engelder. Tracking the burial and tectonic history of Devonian shale of the Appalachian Basin by analysis of joint intersection style. Geol. Soc. Am. Bull., 121 (2009), pp. 265–277

24. D. Lemarchand, J. Gaillardet. Transient features of the erosion of shales in the Mackenzie basin (Canada), evidences from boron isotopes. Earth Planet. Sci. Lett., 245 (2006), pp. 174–189

25. M. McCaffrey, B. Lazar, H. Holland. The evaporation path of seawater and the co-precipitation of Br and K with halite. J. Sed. Petrol., 57 (1987), pp. 928–937

26. S.G. Osborn, A. Vengosh, N.R. Warner, R.B. Jackson. Methane contamination of drinking water accompanying gas-well drilling and hydraulic fracturing. Proc. Nat. Acad. Sci. USA, 108 (2011), pp. 8172–8176

27. S.G. Osborn, A. Vengosh, N.R. Warner, R.B. Jackson. Reply to Saba and Orzechowski and Schon: Methane contamination of drinking water accompanying gas-well drilling and hydraulic fracturing. Proc. Nat. Acad. Sci. USA, 108 (2011), pp. E665–E666

28. Rowan, E., Engle, M., Kirby, C., Kraemer, T., 2011. Radium content of oil- and gas-field produced waters in the northern Appalachian Basin (USA)—Summary and discussion of data. U.S. Geol. Surv. Scient. Invest. Rep. 2011–5135.

29. M. Schoell. The hydrogen and carbon isotopic composition of methane from natural gases of various origins. Geochim. Cosmochim. Acta, 44 (1980), pp. 649–661

30. A.J. Spivack, J.M. Edmond. Boron isotope exchange between seawater and the oceanic crust. Geochim. Cosmochim. Acta, 51 (1987), pp. 1033–1043

31. US-EIA, 2010. Annual Energy Outlook 2010 with Projections to 2035, Washington, DC. <http://www.eia.doe.gov/oiaf/aeo/>.

32. US-EIA, 2011. Review of Emerging Resources: U.S. Shale Gas and Shale Oil Plays. US Dept. of Energy, <http://www.eia.gov/analysis/studies/usshalegas/>.

33. USGS, 2013. National Water Information System Database, <http://waterdata.usgs.gov/nwis> (accessed 01.04.13).

34. A. Vengosh, Y. Kolodny, A. Starinsky, A.R. Chivas, M.T. Mcculloch. Coprecipitation and isotopic fractionation of boron in modern biogenic carbonates. Geochim. Cosmochim. Acta, 55 (1991), pp. 2901–2910

35. N.R. Warner, R.B. Jackson, T.H. Darrah, S.G. Osborn, A. Down, K. Zhao, A. White, A. Vengosh. Geochemical evidence for possible natural migration of Marcellus Formation brine to shallow aquifers in Pennsylvania. Proc. Nat. Acad. Sci. USA (2012) http://dx.doi.org/10.1073/ pnas.1121181109

36. T.R. Weaver, S.K. Frape, J.A. Cherry. Recent cross-formational fluid flow and mixing in the shallow Michigan Basin. Geol. Soc. Am. Bull., 107 (1995), pp. 697–707
37. M.J. Whiticar. Carbon and hydrogen isotope systematics of bacterial formation and oxidation of methane. Chem. Geol., 161 (1999), pp. 291–314
38. M.J. Whiticar, E. Faber, M. Schoell/ Biogenic methane formation in marine and freshwater environments: CO2 reduction vs. acetate fermentation—Isotope evidence. Geochim. Cosmochim. Acta, 50 (1986), pp. 693–709
39. Wilde, F., 2006 Collection of water samples (ver. 2.0): U.S. Geological Survey Techniques of Water-Resources Investigations, Book 9 (Chapter A4).
40. J. Zumberge, K. Ferworn, S. Brown. Isotopic reversal ('rollover') in shale gases produced from the Mississippian Barnett and Fayetteville formations. Mar. Petrol. Geol., 31 (2012), pp. 43–52

Table 1 and several supplemental files are not available in this version of the article. To view this additional information, please use the citation on the first page of this chapter.

CHAPTER 6

Radionuclides in Fracking Wastewater: Managing a Toxic Blend

VALERIE J. BROWN

Naturally occurring radionuclides are widely distributed in the earth's crust, so it's no surprise that mineral and hydrocarbon extraction processes, conventional and unconventional alike, often produce some radioactive waste. [1] Radioactive drilling waste is a form of TENORM (short for "technologically enhanced naturally occurring radioactive material")—that is, naturally occurring radioactive material (NORM) that has been concentrated or otherwise made more available for human exposure through anthropogenic means. [2] Both the rapidity and the extent of the U.S. natural gas drilling boom have brought heightened scrutiny to the issues of radioactive exposure and waste management.

Perhaps nowhere is the question of drilling waste more salient than in Pennsylvania, where gas extraction from the Marcellus Shale using hydraulic fracturing (fracking) made the state the fastest-growing U.S. producer between 2011 and 2012. [3] The Marcellus is known to have high uranium content, says U.S. Geological Survey research geologist Mark Engle. He says concentrations of radium-226—a decay product of ura-

Radionuclides in Fracking Wastewater: Managing a Toxic Blend. Brown VJ. Environmental Health Perspectives *122,2 (2014), DOI:10.1289/ehp.122-A50. Reproduced from Environmental Health Perspectives*

nium—can exceed 10,000 picocuries per liter (pCi/L) in the concentrated brine trapped in the shale's depths.

To date the drilling industry and regulators have considered the risk posed to workers and the public by radioactive waste to be minor. In Pennsylvania, Lisa Kasianowitz, an information specialist with the state Department of Environmental Protection (PADEP), says there is currently nothing to "indicate the public or workers face any health risk from exposure to radiation from these materials." But given the wide gaps in the data, this is cold comfort to many in the public health community.

6.1 WASTE PRODUCTION AND STORAGE

After fracking, both gas and liquids—including the injected water and any water residing in the formation (known as "flowback" and "produced water" [4])—are pulled to the surface. Fluids trapped in the shale are remnants of ancient seawater. The salts in shale waters reached extreme concentrations over millions of years, and their chemical interactions with the surrounding rock can mobilize radionuclides. [5,6] Several studies indicate that, generally speaking, the saltier the water, the more radioactive it is. [5,7]

Dissolved compounds often precipitate out of the water, building up as radionuclide-rich "scale" inside pipes. To remove the pipe-clogging scale, operators might inject chemicals to dissolve it. [8] Scale also may be removed mechanically using drills, explosives, or jets of fluid, [9] in which case it joins the solid waste stream.

Wastes are often stored temporarily in containers or in surface impoundments, also called pits and ponds. Data on how many such ponds are used in shale gas extraction are sparse, but according to Kasianowitz, there are 25 centralized impoundments in Pennsylvania. Centralized impoundments can be the size of a football field and hold at least 10 million gallons of liquid. Although at any given time the number of smaller ponds is probably much higher, she says these ephemeral lagoons are used mostly in the early phase of well development and are rapidly decommissioned.

Most impoundments are lined with plastic sheeting. Pennsylvania requires that pit liners for temporary impoundments and disposal have a minimum thickness of 30 mil and that seams be sealed to prevent leakage. [10] Ohio's only requirement is that pits must be "liquid tight." [10] However, improper liners can tear, [7] and there have been reports of pit liners tearing and pits overflowing in Pennsylvania and elsewhere. [11]

A small 2013 study of reserve pits in the Barnett Shale region of Texas suggested another consideration in assessing pit safety. Investigators measured radium—the radionuclide generally used as a proxy to judge whether NORM waste complies with regulatory guidelines for disposal—as well as seven other radionuclides not routinely tested for. Although individual radionuclides were within existing regulatory guidelines, total beta radiation in one sample was more than 8 times the regulatory limit. "Evaluating the single radionuclide radium as regulatory exposure guidelines indicate, rather than considering all radionuclides, may indeed underestimate the potential for radiation exposure to workers, the general public, and the environment," the authors wrote. [2]

6.2 SURFACE WATERS

Ultimately most wastewater is either treated and reused or sent to Class II injection wells (disposal or enhanced recovery wells). A small fraction of Pennsylvania's fracking wastewater is still being treated and released to surface waters until treatment facilities' permits come up for renewal under new, more stringent treatment standards, Kasianowitz says.

Concerns about NORM in the Marcellus have recently focused on surface waters in Pennsylvania. That's because until 2011, most produced water was sent to commercial or public wastewater treatment plants before being discharged into rivers and streams, many of which also serve as drinking water supplies. In April of that year PADEP asked all Marcellus Shale fracking operations to stop sending their wastewater to treatment plants, according to Kasianowitz. Although voluntary, this request motivated most producers to begin directly reusing a major fraction of their produced water or reusing it after treatment in dedicated commercial treatment plants that are equipped to handle its contaminants.

A team of Duke University researchers led by geochemist Avner Vengosh sought to characterize the effluent being discharged from one such plant, the Josephine Brine Treatment Facility in southwestern Pennsylvania. The researchers compared radioactivity and dissolved solids in sediment both up- and downstream of the facility and found a 90% reduction in radioactivity in the effluent. The radioactive constituents didn't just disappear; the authors noted that most had likely been transferred and accumulated to high levels in the sludge that would go to a landfill. [12]

Stream sediments at the discharge site also had high levels of radioactivity, keeping it out of the surface water downstream but posing the risk of bioaccumulation in the local food web. The outflow sediment radiation levels at the discharge site were 200 times those in upstream sediments. The study highlighted "the potential of radium accumulation in stream and pond sediments in many other sites where fracking fluids are accidentally released to the environment," says Vengosh.

The study also demonstrated another potential impact of treated brine on water quality. Most produced water contains bromide, which can combine with naturally occurring organic matter and chlorine disinfectant to form drinking water contaminants called trihalomethanes. These compounds are associated with liver, kidney, and nervous system problems. [13] The Duke researchers reported highly elevated concentrations of bromide over a mile downstream from the plant—a potential future burden for drinking water treatment facilities downstream. [12]

6.3 DEEP INJECTION

Following the 2011 policy change, Ohio's Class II injection wells began to receive much of Pennsylvania's end-stage wastewater. Pennsylvania's geology does not lend itself to this method; the state has only six injection wells available for this purpose, while Ohio has 177, [10] and Texas has 50,000. [14]

Class II injection wells place the wastewater below the rock strata containing usable groundwater. Conventional industry wisdom says this prevents migration of contaminants into shallower freshwater zones. [7,15,16,17]

But some believe this may be a flawed assumption. The reason fracking works to force gas out of the rock is also why some observers think injection wells could be unstable—the extreme pressure of injection can take nearly a year to dissipate, according to hydrologic consultant Tom Myers, who published a modeling study of fracking fluids' underground behavior in 2012. [18]

Myers says the lingering higher-than-normal pressure could bring formation waters, along with fracking chemicals, closer to the surface far faster than would occur over natural geological time scales of thousands of years. This is particularly true if there are faults and/or abandoned wells within the fracking zone.

Another study has demonstrated the possibility that formation water can migrate into freshwater aquifers through naturally occurring pathways. [19] Although the pathways were not, themselves, caused by gas drilling, the study authors suggest such features could make certain areas more vulnerable to contamination due to fracking.

Asked about the integrity of deep-injection wells, Vengosh says, "As far as I know nobody's actually checking." If such leaks were happening, he says, much would depend on how they connected to drinking water aquifers. "Unlike freshwater systems where radium would accumulate in the sediments," he says, "if you have a condition of high salinity and reducing conditions, radium will be dissolving in the water and move with the water."

6.4 BENEFICIAL USES AND LANDFILLS

Fracking wastes may also be disposed of through "beneficial uses," which can include applying produced water as a road de-icer or dust suppressant, using drilling cuttings in road maintenance, and spreading liquids or sludge on fields. [12,20,21] Pennsylvania allows fracking brine to be used for road dust and ice control under a state permit. [22] While the permit sets allowable limits for numerous constituents, radioactivity is not included. [23]

Conventional wisdom about radium's stability in landfills rests on an assumption regarding its interaction with barite (barium sulfate), a common constituent in drilling waste. However, Charles Swann of the Mississippi Mineral Resources Institute and colleagues found evidence that radium in waste spread on fields may behave differently in soil than expected. When they mixed scale comprising radium and barite with typical Mississippi soil samples in the laboratory, radium was gradually solubilized from the barite, probably as a result of soil microbial activity. "This result," the authors wrote, "suggests that the landspreading means of scale disposal should be reviewed." [24]

Solids and sludges can also go to landfills. Radioactivity limits for municipal landfills are set by states, and range from 5 to 50 pCi/g. [25] Since Pennsylvania began requiring radiation monitors at municipal landfills in 2001, says Kasianowitz, fracking sludges and solids have rarely set them off. In 2012 they accounted for only 0.5% of all monitor alarms. They "did not contain levels of radioactivity that would be acutely harmful to the public," according to a 2012 review of Pennsylvania's fracking practices by the nonprofit State Review of Oil and Natural Gas Environmental Regulations. [26] Dave Allard, director of PADEP's Bureau of Radiation Protection, points out that because all soils contain at least some radionuclides, "you're always going to have some radium, thorium, and uranium, because these landfills are in soils."

6.5 ASSESSING EXPOSURES

At the federal level, radioactive oil and gas waste is exempt from nearly all the regulatory processes the general public might expect would govern it. Neither the Atomic Energy Act of 1954 nor the Low-Level Radioactive Waste Policy Act covers NORM. [2] The Nuclear Regulatory Commission has no authority over radioactive oil and gas waste. State laws are a patchwork. Workers are covered by some federal radiation protections, although a 1989 safety bulletin from the Occupational Safety and Health Administration noted that NORM sources of exposure "may have been overlooked by Federal and State agencies in the past." [27]

Fracking in the Marcellus has advanced so quickly that public understanding and research on its radioactive consequences have lagged behind, and there are many questions about the extent and magnitude of the risk to human health. "We are troubled by people drinking water that [could potentially have] radium-226 in it," says David Brown, a public health toxicologist with the Southwest Pennsylvania Environmental Health Project. "When somebody calls us and says 'is it safe to drink our water,' the answer is 'I don't know.'"

PADEP is conducting a study to determine the extent of potential exposures to radioactive fracking wastewater. [28] The PADEP study will sample drill cuttings, produced waters, muds, wastewater recycling and treatment sludges, filter screens, extracted natural gas, scale buildup in well casings and pipelines, and waste transport equipment. PADEP will also evaluate radioactivity at well pads, wastewater treatment plants, wastewater recycling facilities, and landfills.

The EPA is studying the issue with a review of the potential impacts of hydraulic fracturing, [29] including radioactivity, on drinking water resources. A draft of the EPA study will be released for public comment and peer review in late 2014, according to Christopher Impellitteri, chief of the Water Quality Management Branch at the agency's National Risk Management Research Laboratory.

The EPA study includes research designed to assess the potential impacts from surface spills, well injection, and discharge of treated fracking wastewater on drinking water sources. One project will model the transport of contaminants, including radium, from treatment outflows in receiving waters. Field and laboratory experiments will characterize the fate and transport of contaminants in wastewater treatment and reuse processes. Groundwater samples are being tested for radium-226, radium-228, and gross alpha and beta radiation. The overall study does not include radon. [29]

Both radon and radium emit alpha particles, which are most dangerous when inhaled or ingested. When inhaled, radon can cause lung cancer, and there is some evidence it may cause other cancers such as leukemia. [30] Consuming radium in drinking water can cause lymphoma, bone cancer, and leukemias. [31] Radium also emits gamma rays, which raise cancer risk throughout the body from external exposures. Radium-226 and radium-228 have half-lives of 1,600 years and 5.75 years, respectively. Radium is known to bioaccumulate in invertebrates, mollusks, and freshwater fish, [12] where it can substitute for calcium in bones. Radium eventually decays to radon; radon-222 has a half-life of 3.8 days.

Geochemically, radon and radium behave differently. Radon is an inert gas, so it doesn't react with other elements and usually separates from produced water along with methane at the wellhead. Although there are few empirical data available, the natural gas industry has not been concerned about radon reaching its consumers in significant amounts, in part because of radon's short half-life and because much of it is released to the atmosphere at the wellhead. [32]

6.6 BEYOND ASSUMPTIONS

Assumptions about quality control underlie much of the debate about whether the risks of fracking outweigh the benefits. "If everything is done the way it's supposed to be done, the impact of this radioactivity would be fairly minimal in the environment in Pennsylvania, because they're reusing the water," says Radisav R. Vidic, a professor of civil and environmental engineering at the University of Pittsburgh. "The only potential pathway is an accident, a spill, or a leak." But, he adds, "That's something that happens in every industry, so there's nothing you can do about it."

Indeed, Vengosh says, PADEP has reports of hundreds of cases of spills and contamination that involved fracking fluids. Furthermore, he says, "The notion that the industry can reuse all flowback and produced water is simply not possible, given the chemistry of the wastewater."

Many of the studies to date on fracking's environmental impacts have suffered from a lack of access to actual treatment practices, according to Engle. He attributes this to a lack of trust between the industry and scientists, and the fact that such information is often proprietary. But Swann reports a different experience working with Mississippi producers. "The small, independent producers were very willing to cooperate and gladly provided assistance, often at their expense," he says. "Only through their assistance were we able to sample so many fields and wells." [24]

Research published in December 2013 suggests one potential new treatment for radioactivity in fracking waste. [33] Vengosh and colleagues combined various proportions of flowback water with acid mine drainage (AMD) to test the possibility of using the latter as an alternative source of water for fracking. AMD—acidic leachate from mining sites and other disturbed areas—is an important water pollutant in some regions. Laboratory experiments showed that mixing flowback water with AMD caused much of the NORM in the flowback to precipitate out, leaving water with radium levels close to EPA drinking water standards.

The authors suggest the radioactive precipitate could be diluted with nonradioactive waste to levels appropriate for disposal in municipal land-fills. If it can be brought to industrial scale, Vengosh says, this method could provide a beneficial use for AMD while reducing the need for freshwater in fracking operations and managing the inevitable radioactive waste.

Studies such as this provide a light at the end of the wellbore. Yet the current patchy understanding of radioactive fracking waste's fate in the environment precludes making good decisions about its management. And even if fracking the Marcellus ceased overnight, the questions and potential problems about radioactivity would linger. "Once you have a re-lease of fracking fluid into the environment, you end up with a radioactive legacy," says Vengosh.

REFERENCES

1. EPA. TENORM: Oil and Gas Production Wastes [website]. Washington, DC:Office of Radiation and Indoor Air, U.S. Environmental Protection Agency (updated 30 August 2012). Available: http://www.epa.gov/radiation/tenorm/oilandgas.html [accessed 8 January 2014].
2. Rich AL, Crosby EC. Analysis of reserve pit sludge from unconventional natural gas hydraulic fracturing and drilling operations for the presence of technologically en-hanced naturally occurring radioactive material (TENORM). New Solut 23(1):117–135 (2013); http://dx.doi.org/10.2190/NS.23.1.h.
3. EIA. Pennsylvania is the fastest-growing natural gas-producing state [weblog entry]. Today in Energy (17 December 2013). Washington, DC:U.S. Energy Information Administration, U.S. Department of Energy. Available: http://www.eia.gov/today-inenergy/detail.cfm?id=14231# [accessed 8 January 2014].
4. Schramm E. What Is Flowback and How Does It Differ from Produced Water? [web-site]. Wilkes-Barre, PA:The Institute for Energy and Environmental Research for Northeastern Pennsylvania, Wilkes University (24 March 11). Available: http://en-ergy.wilkes.edu/pages/205.asp [8 January 2014].
5. Rowan EL, et al. Radium Content of Oil- and Gas-Field Produced Waters in the Northern Appalachian Basin (USA)—Summary and Discussion of Data. Scientific Investigations Report 2011–5135. Washington, DC:U.S. Geological Survey, U.S. Department of the Interior (2011). Available: http://pubs.usgs.gov/sir/2011/5135/pdf/sir2011-5135.pdf [accessed 8 January 2014.].

6.   Haluszczak LO, et al. Geochemical evaluation of flowback brine from Marcellus gas wells in Pennsylvania, USA. Appl Geochem 28:55–61 (2012); http://dx.doi.org/10.1016/j.apgeochem.2012.10.002.

7.   GAO. Oil and Gas: Information on Shale Resources, Development, and Environmental and Public Health Risks. GAO-12-732. Washington, DC:U.S. Government Accountability Office (5 September 2012). Available: http://www.gao.gov/products/GAO-12-732 [accessed 8 January 2014].

8.   IAEA. Radiation Protection and the Management of Radioactive Waste in the Oil and Gas Industry, Safety Reports Series No. 34. Vienna, Austria:International Atomic Energy Agency (November 2003). Available: http://www-pub.iaea.org/MTCD/publications/PDF/Pub1171_web.pdf [accessed 8 January 2014].

9.   Crabtree M, et al. Fighting scale: removal and prevention. Oilfield Rev 11(3):30–45 (1999); http://goo.gl/AAyqxE.

10. GAO. Unconventional Oil and Gas Development: Key Environmental and Public Health Requirements. GAO-12-874. Washington, DC:U.S. Government Accountability Office (September 2012). Available: http://www.gao.gov/assets/650/647782.pdf [accessed 8 January 2014].

11. Legere L. Hazards posed by natural gas drilling not always underground. Scranton Times-Tribune, News section, online edition (21 June 2010). Available: http://goo.gl/L0k9OP [accessed 8 January 2014].

12. Warner NR, et al. Impacts of shale gas wastewater disposal on water quality in western Pennsylvania. Environ Sci Technol 47(20):11849–11857 (2013); http://dx.doi.org/10.1021/es402165b.

13. EPA. Drinking Water Contaminants: Basic Information about Disinfection Byproducts in Drinking Water: Total Trihalomethanes, Haloacetic Acids, Bromate, and Chlorite [website]. Washington, DC:Office of Water, U.S. Environmental Protection Agency (updated 13 December 2013). Available: http://goo.gl/hnYJbj [accessed 8 January 2014].

14. Clark CE, Veil JA. Produced Water Volumes and Management Practices in the United States. ANL/EVS/R-09/1. Argonne, IL:Environmental Science Division, Argonne National Laboratory (September 2009).

15. U.S. Environmental Protection Agency. Underground Injection Control Program: Criteria and Standards. 40CFR Part 146, Subpart C, Section 146.22: Construction Requirements. Washington, DC:U.S. Government Printing Office (1 July 2012). Available: http://goo.gl/1U7fjX [accessed 8 January 2014].

16. Flewelling SA, Sharma M. Constraints on upward migration of hydraulic fracturing fluid and brine. Ground Water 52(1):9–19 (2013); http://dx.doi.org/10.1111/gwat.12095.

17. Jackson RE, et al. Ground protection and unconventional gas extraction: the critical need for field-based hydrogeological research. Ground Water 51(4):488–510 (2013); http://www.ncbi.nlm.nih.gov/pubmed/23745972.

18. Myers T. Potential contaminant pathways from hydraulically fractured shale to aquifers. Ground Water 50(6):872–882 (2012); http://dx.doi.org/10.1111/j.1745-6584.2012.00933.x.

19. Warner NR, et al. Geochemical evidence for possible natural migration of Marcellus Formation brine to shallow aquifers in Pennsylvania. Proc Natl Acad Sci USA 109(30):11961–11966 (2012); http://dx.doi.org/10.1073/pnas.1121181109.

20. Drilling Waste Management Information System. Fact Sheet—Beneficial Reuse of Drilling Wastes [website]. Argonne, IL:Argonne National Laboratory (undated). Available: http://web.ead.anl.gov/dwm/techdesc/reuse/ [accessed 8 January 2014.].

21. Guerra K, et al. Oil and Gas Produced Water Management and Beneficial Use in the Western United States. Science and Technology Program Report No. 157. Denver, CO:Bureau of Reclamation, U.S. Department of the Interior (September 2011). Available: http://www.usbr.gov/research/AWT/reportpdfs/report157.pdf [accessed 8 January 2014].

22. Poole H. State Policies on Use of Hydraulic Fracturing Waste as a Road Deicer. Hartford, CT:Office of Legislative Research, Connecticut General Assembly (undated). Available: http://www.cga.ct.gov/2013/rpt/2013-R-0469.htm [accessed 8 January 2014].

23. PADEP. Special Conditions General Permit WMGR064. Harrisburg, PA:Pennsylvania Department of Environmental Protection, Commonwealth of Pennsylvania (undated). Available: http://www.portal.state.pa.us/portal/server.pt?open=18&objID=505511&mode=2 [accessed 8 January 2014].

24. Swann C, et al. Evaluations of Radionuclides of Uranium, Thorium, and Radium Associated with Produced Fluids, Precipitates, and Sludges from Oil, Gas, and Oilfield Brine Injections Wells in Mississippi, Final Report. University, MS:Mississippi Mineral Resources Institute/Department of Pharmacology/Department of Geology and Geological Engineering, University of Mississippi (March 2004). Available: http://www.olemiss.edu/depts/mmri/programs/norm_final.pdf [accessed 8 January 2014].

25. Walter GR, et al. Effect of biogas generation on radon emissions from landfills receiving radium-bearing waste from shale gas development. J Air Waste Manag Assoc 62(9):1040–1049 (2012); http://dx.doi.org/10.1080/10962247.2012.696084.

26. STRONGER. Pennsylvania Follow-up State Review. Middletown, PA:State Review of Oil and Natural Gas Environmental Regulations, Inc. (September 2013). Available: http://goo.gl/DQhOlA [accessed 8 January 2014].

27. OSHA. Health Hazard Information Bulletin: Potential Health Hazards Associated with Handling Pipe Used in Oil and Gas Production. Washington, DC:Occupational Safety & Health Administration, U.S. Department of Labor (26 January 1989). Available: https://www.osha.gov/dts/hib/hib_data/hib19890126.html [accessed 8 January 2014].

28. PADEP. Oil & Gas Development Radiation Study [website]. Harrisburg, PA:Pennsylvania Department of Environmental Protection, Commonwealth of Pennsylvania (2014). Available: http://goo.gl/P22FQM [accessed 8 January 2014].

29. EPA. The Potential Impacts of Hydraulic Fracturing on Drinking Water Resources: Progress Report. EPA 601/R-12/011. Washington, DC:Office of Research and Development, U.S. Environmental Protection Agency (December 2012). Available: http://goo.gl/4YfBka [accessed 8 January 2014].

30. NRC. Health effects of radon progeny on non-lung-cancer outcomes. In: Health Effects of Exposure to Radon, BEIR VI. Washington, DC:Committee on Health Risks of Exposure to Radon (BEIR VI), National Research Council, National Academies Press (1999). Available: http://www.nap.edu/openbook.php?record_i d=5499&page=118 [accessed 8 January 2014].
31. EPA. Radionuclides: Radium [website]. Washington, DC:Office of Radiation and Indoor Air, U.S. Environmental Protection Agency (updated 6 March 2012). Available: http://www.epa.gov/radiation/radionuclides/radium.html#affecthealth [accessed 8 January 2014].
32. EPA. Radioactive Waste from Oil and Gas Drilling. EPA 402-F-06-038. Washington, DC:Office of Radiation and Indoor Air, U.S. Environmental Protection Agency (April 2006). Available: http://www.epa.gov/rpdweb01/docs/drilling-waste.pdf [accessed 8 January 2014].
33. Kondash AJ, et al. Radium and barium removal through blending hydraulic fracturing fluids with acid mine drainage. Environ Sci Technol 48(2):1334–1342 (2014); http://dx.doi.org/10.1021/es403852h.

PART III

THE QUEST FOR MITIGATION

Optimal Well Design for Enhanced Stimulation Fluids Recovery and Flowback Treatment in the Marcellus Shale Gas Development using Integrated Technologies

RICHARD OLAWOYIN, CHRISTIAN MADU, AND KHALED ENAB

7.1 INTRODUCTION

The advent of intensified exploration and production in the Marcellus Shale has brought benefits and challenges to the area. An enormous investment is being made in developing extraction technologies to effectively recover natural gas from tight formations that is very low permeability [1]. Directional drilling technologies alongside hydraulic fracturing have made formations once considered uneconomical, an investor's hub. The combination of directional drilling and hydraulic fracturing is quintessential in accessing the pay zone efficiently and also creating a network of fractures in the formation, which enables an optimal recovery of the reservoir fluid [2,3]. Wells are enhanced considerably using the hydraulic fracturing method, especially in the Marcellus Shale with an estimated 516 trillion

Optimal Well Design for Enhanced Stimulation Fluids Recovery and Flowback Treatment in the Marcellus Shale Gas Development using Integrated Technologies. © *Olawoyin R, Madu C, and Enab K. Hydrology:* Current Research *3,141 (2012), doi:10.4172/2157-7587.1000141. Licensed under a Creative Commons Attribution License, http://creativecommons.org/licenses/by/3.0/.*

cubic feet (TCF) of gas in place [4]. This capacity makes the Marcellus Shale economically important to the world's strategic energy prospects. Several analyst have made different estimates of the gas reserves in the Marcellus Shale, with new studies, more accurate estimates are possible in the near future. At present the formation is estimated to contain approximately 1,307 TCF [5] Activities in the region have increased exponentially, triggering a queue of investors eager to take part in the booming energy venture. Towards this end, there has been a huge capital intensive investment in the Marcellus Shale by some international energy firms in developing productive fields, providing advanced technology and in the exploration of the natural resource of interest (gas). Details on the activities involving shale gas developments have been resented in other literatures [6-14]. The waste generated due to increasing exploration remains a problem [15-17], there are constraints in form of regulations that demand that the waste water be treated before re injected into the natural flow. This paper presents a reservoir simulation identifying the most prolific drilling design and also provides cost effective solution to tackling the problem of water usage for hydraulic fracturing purposes, load recovery (flow-back), waste water treatment and energy consumption.

7.2 BACKGROUND ANALYSIS: RESERVOIR SIMULATION

With the goal of reducing the impact of drilling in the Marcellus shale area, several directional wells were designed using CMG to identify the most suitable well design (within a 640 acres area) with the least environmental impact and highest recovery. The reservoir properties used was from one of the most prolific areas in the Marcellus Shale, in the Bradford County (Table 1).

7.2.1 SINGLE HORIZONTAL WELL

One horizontal well (1440ft) was designed with 8710 ft total depth (TD) (Figure1) from the surface and deviated depth (DD) of 90ft vertically into the reservoir. For a 40 year CMG pressure profile, the pressure distribution as represented in figure 1b shows that the steady production or gas recovery begins at (3.5E+06) ft³/day as shown in figure 2a, with cumulative gas production of 6.2E+09 ft³ (Figure 2b). This model suggests that though substantial amount of gas can be produced but a single well is insufficient for optimal production and footprint reduction.

TABLE 1: Typical reservoir properties for the designed well.

Depth	8620 Ft
Thickness	180 Ft
i	5280 Ft
j	5280 Ft
Permeability (horizontal)	0.01 md
Permeability (vertical)	0.001
Porosity	5%
Temperature	210 F
Initial Pressure	4000 psi
Rock compressibility	0.000001

7.2.2 MULTILATERAL HORIZONTAL WELLS

Simulating a dual horizontal well with the same TD and DD as the single horizontal well, the gas rate will be 5.1E+06 ft³/day and the cumulative gas production of 9.7E+09 ft³ as seen in figures 3a and 3b respectively. Illustrated in figures 4a and 4b are the reservoir pressure profile, or the extent of pressure disturbance, around the well bore after a period of 40 years and the well designs respectively. The profile shows steady production indicating that a dual horizontal well gives better results in both design and production profile than the single horizontal well. Other designs considered in this studyinclude;the dual horizontal wells with distance, as shown in figures 5a and 5b,which shows the gas production rate and cumulative gas production over a period of 40 years. The two horizontal wells were designed at opposing angles and at variable depths (30 ft apart) both wells are within the pay zone. The production profile is enhanced with more gas produced than the single and conventional dual horizontal wells.

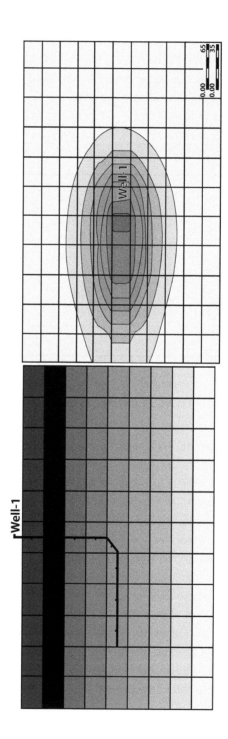

FIGURE 1: a) Horizontal well designed in the Marcellus Shale. b) Pressure distribution for 1 horizontal well: 40 years profile.

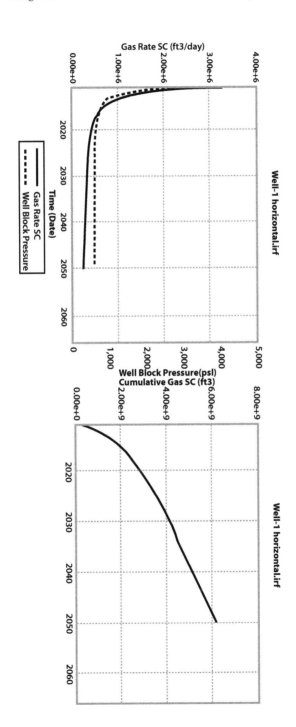

FIGURE 2: a) Gas production rate vs. Time for a single horizontal well. b) Cumulative gas production vs. time for a single horizontal well.

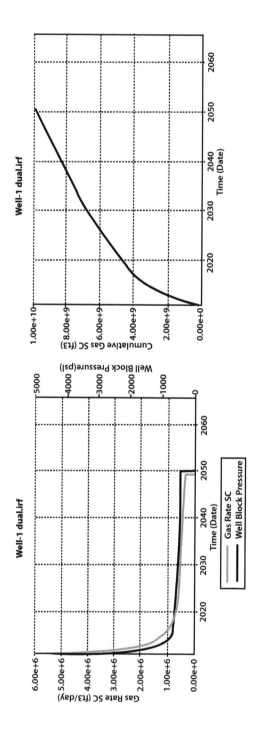

FIGURE 3: a) Gas production rate vs. time for dual horizontal well. b) Cumulative gas production vs. time for dual horizontal well.

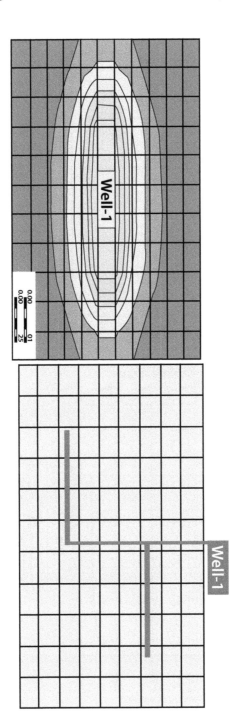

FIGURE 4: a) Pressure distribution for dual horizontal well: 40 years profile. b) Dual horizontal well with distance.

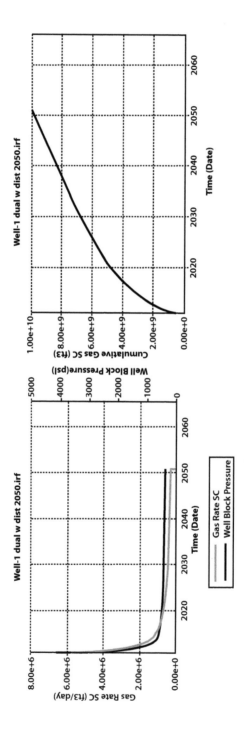

FIGURE 5: a) Gas production rate vs. time for dual horizontal wells with distance. b) Cumulative gas vs. time for dual horizontal wells with distance.

Multilateral wells pressure distribution profile and design as shown in figures 6a and 6b respectively illustrate a wider coverage of the reservoir. The gas production rate begins at 1.18 E+07 ft³/day as shown in figure 7a, while the cumulative gas production was estimated as 2.6 E+10 ft³ as shown in figure 7b. There is a significant improvement in the flow rate and cumulative recovery rate from the multilateral well simulation design with higher productivity. The main objective is to identify the best well design that will require the least number of surface wells and also have the capability to drain the same area effectively as shown in the comparative cumulative gas production profile for all the well designs (Figure 8). Selecting the multilateral well design will reduce surface disturbance, while the entire reservoir is effectively drained over time. In this study, we have presented the optimal well design for the purpose of understanding the fluid requirements for the well drilling and completion.

7.3 ENVIRONMENTAL MANAGEMENT OF PROCESS FLUIDS

The use of silicate drilling fluid gives good results as water based drilling mud (WBM) due to the high penetration rate, most favorable inhibition characteristics, and high solids discharge performance, less trouble time, least environmental impact and better wellbore integrity. Bentonite cement gives good results in the Marcellus shale since it reduces slurry cost and facilitates in the decrease in the cement thickening time [18].

7.3.1 STIMULATION TECHNIQUES AND FLUID RECOVERY

Well stimulation methods have progressed with the development of unconventional reservoirs due to the profitable possibility of unconventional field exploration. Success depends mostly on efficient stimulation of rocks with permeability ranging from 10 to 100 nanodarcies [19]. The potential for Shale gas has become very important in the energy market. The stimulation process has continued to improve in time through advanced innovation [3]. Conversely, the most feasible technique employed in stimulating the well (hydraulic fracturing method) has aroused a lot of attention since its use became predominant in the region. As part of the process, the technique generates huge amount of chemically contaminated water from the sub surface. Hence, it has become imperative to look for ways in curtailing the utilization of water resource and also develop better methods of flow back recovery.

7.3.2 THE HYDRAULIC FRACTURING PROCESS

Hydraulic fracturing entails the perforation of the shale rock at depth with a perforating gun and subsequently injecting high pressured water inside the well bore, this then generate set of fractures in the rock shale and widens fractures present prior to the stimulation. The prime function of well stimulation in shale reservoirs is to expand the radius of the drainage by generating lengthy fractures that intersect natural fractures thus setting up a flow channel network for the gas towards the wellbore, by which the stimulated reservoir volume (SRV) is maximized [20]. Figure 9 shows a typical fracture design profile for the Marcellus Shale a Bradford County, Pennsylvania.

This increases the permeability and enhances production from the formation. This process is a function of pressure and permeability, as fluid is pumped into the formation at high pressure, the blocks nearest to the well bore experiences breakage and fractures are created, which improves the permeability. Slick water (approximately 98% water) is the preferred fluid for stimulation in low-permeability reservoirs, and it is also the principal tool in breaching tight formations in unconventional plays provided there is viable water accessibility [21]. Precipitation in the area (Appalachian) is approximately 43 inches per year, compared to the other parts of the continental United States that receive about 10 inches [22]. Additionally, there are other several available water sources which may include, municipal water, rivers, ponds, lakes cited close to the Marcellus shale area, therefore the use of slick water for the hydraulic fracture process has proven to be the most cost effective and convenient At the end of the hydraulic fracturing well stimulation treatment, some of the injected fluids in the formation will flow into the well and to the well head, since the pumping pressure has been eliminated [23]. It is important for this flow back water to be recovered because the continuous presence of the fluid would block the free passage of the natural gas through the propped fractures. The composition of the fracturing fluid changes once it has had contact with the reservoir rocks, due to contamination and dissolution of several other minerals in the formation. It has been estimated that in the Marcellus Shale area, 35% of the fracturing fluid is recovered as flow back fluids. This is not economically beneficial to operators since more volume of fresh water would be required for successive stimulation stages [24]. In other Shale plays however, the recovery rate has been estimated to be between 30%-70% [23]. Consequently, increasing the recovery rate of flow back fluids would minimize the need for more fresh water and also reduce the general impact of shale gas drilling in the area. A load recovery of about 86 % of the pump in fluid is attainable with the fracture design which amount to 73,714 bbl.

7.3.3 NATURE OF FLOW BACK WATER

Several reviews have attempted to capture the mineralogy, composition, environmental impact and existing practices related to the use and the management of flow back fluids in unconventional and conventional gas fields [25].

Total dissolved solids (TDS) which is the amount of soluble salts flow back fluid varies from about 20 mg/l to about 200,000 mg/l, representing a salt content of above 20% in the Marcellus Shale. The salt content at locations is dependent on several factors ranging from the composition of the formation to variations to natural conditions in the formation and also the tendency for the fracturing fluid to flow through the formation.

The level of dissolved components rises while the pH and alkalinity decreases as flow back advanced. Sodium and calcium exhibit analogous patterns in the developed wells. This is prone to sulfate scaling as the amount of calcium rises while sulfate drops. A sharp increase in barium concentrations in the late phases of flow back which amount to load recovery of 30% indicate the possibility of scaling from barium sulfate development as the flow back progresses.

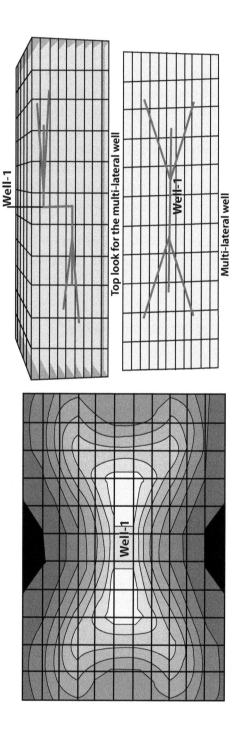

FIGURE 6: a) Pressure distribution for multilateral well: 40 years profile. b) Multilateral wells illustration.

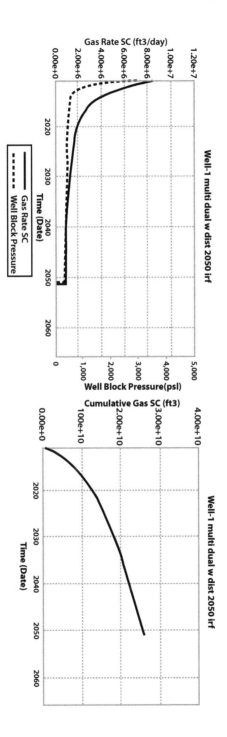

FIGURE 7: a) Gas production rate vs. time for well #1 of multilateral (40years). b) Cumulative gas vs. production time for multilateral wells.

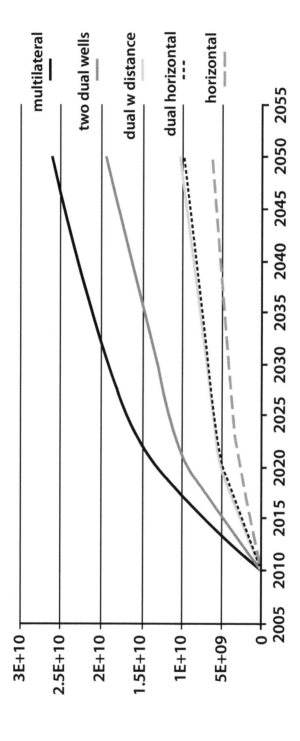

FIGURE 8: Cumulative gas production for all well designs: 40 years profile.

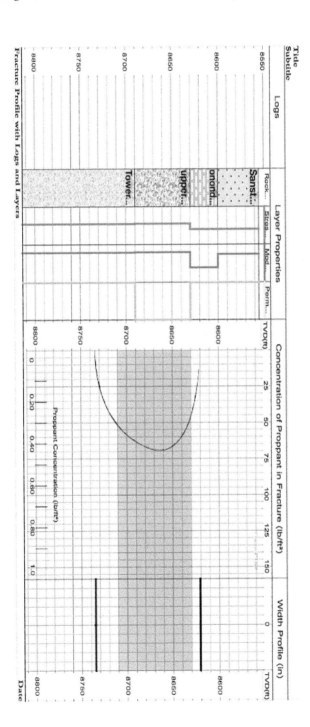

FIGURE 9: Fracture profile in the Marcellus Shale Bradford County.

TABLE 2: Marcellus Shale Well Flowback Analysis Data for 20 days.

Samp Date	TDS	Ca (mg/L)	Mg (mg/L)	CaCO₃	Na (mg/L)	K (mg/L)	Fe (mg/L)	Ba (mg/L)	Sr (mg/L)	Mn (mg/L)	SO₄ (mg/L)	Cl (mg/L)
4/14	22438	15.00	2.73	49.44	18.00	1.65	0.25	0.23	0.46	0.60	3.00	183.00
4/26	84839	7100	603	23286	22800	326	3.93	2000	1400	6.69	0.00	50600
4/27	89861	7640	651	24952	24300	346	7.80	1990	1510	7.07	8.87	53400
4/27	105161	8490	714	27432	25100	352	9.70	1870	1670	7.44	156	66800
4/28	116266	10500	893	33879	29400	410	35.30	1980	2200	9.10	139	70700
4/29	123902	11700	996	38419	31100	437	16.20	2480	2860	9.50	2.94	74300
4/30	164081	16700	1400	52071	41700	579	23.50	2230	2570	13.00	165	98700
5/1	140169	14000	1150	44358	34300	477	28.70	2290	2590	11.00	22.70	85300
5/2	146539	16700	1380	53473	39400	535	30.20	3000	3380	13.10	0.19	82100
5/3	161636	17100	1410	54446	40400	543	35.20	2950	3280	13.30	4.97	95900
5/4	164902	16700	13000	103026	37000	496	32.90	3850	4310	12.30	1.15	89500

Table 2 shows data from a Marcellus Shale Well in Bradford County flow back fluids analyzed in a 20 day period. Barium is significantly high; reaching a value of 3.1 mg/L. The highest concentration of Strontium (Sr) was recorded at a value of 4.3 g/L on the twentieth day. It was observed that Sr levels reached a high value of 15 g/L in flow back at other locations in the Marcellus Shale wells. Potassium concentrations area about 5 g/L at 20 days.

As the chemical composition of Marcellus flow back water varies dependent upon the well location and elapsed time since the fracture was completed, the results in table 1 are typical of flow back recoveries in the Marcellus Shale. The concentration of brine in the flow back increases with time and reaches the optimum, at this point; the flow back will have a steady decline in the concentration of brine as the well production continues.

7.4 METHOD AND DESIGN: FORWARD OSMOSIS

Forward Osmosis (FO) is a distinctive and developing technology which treats waste water and necessitates no energy to pressure the flow through the membrane system, thereby eliminating the need for huge energy requirement. A draw solution is employed across the alternate side of the membrane to drive a high osmotic pressure generating the pressure gradient than the other side of the membrane which contains the waste stream [26].

It is essential that the solute such as sodium chloride (NaCl), used as the draw solution is usable with the waste water, in other to generate pure brine. Ammonium (NH_3)-carbon dioxide (CO_2) gas mixture can also be used. It can be removed from the solution comparatively easily [27]. The FO system functions effectively without the need of an applied hydraulic pressure. Other advantages include the low tendency for membrane fouling compared to the Reverse Osmosis (RO) systems, and a high rate of contaminants rejection [28,29]. It has been proven that it removes 97-99 percent of salts and heavy metals from waste water and it is also effective in rejecting viruses, bacteria and other colloidal solids with a 100 percent success rate [30].

The rate of energy that facilitates water flow, P is shown in equation 1.

$$P = q\Delta\Pi \tag{1}$$

Water flux through a membrane is represented as q, while the pressure difference on the membrane is represented as $\Delta\pi$ [29]. The pressure difference can be further defined as the difference in the draw solution's osmotic energy and that of the feed. This is illustrated in equation 2 [29].

$$\Delta\Pi = \Pi_{DS} - \Pi_{feed} \tag{2}$$

Π_{DS} is the draw solution osmotic pressure, measured in Pascal (Pa) while Π_{feed} is feed's osmotic pressure (of water) in Pa. Equation 3 explains the water passage in the osmotic process.

$$J_W = A(\sigma\Delta\Pi - \Delta P) \tag{3}$$

J_W is the flux of water, while A stands for the membrane water permeability, σ represents reflection coefficient, the applied pressure is represented as ΔP [31].

7.4.1 DRAW SOLUTION

The Forward Osmosis process does not require to be pressured in order for the flow to be activated. Instead on the permeate section of the membrane, concentrated draw solution drive the flow due to concentration difference [32]. The application of this design process has been extensively studied and reported in related reports [27]; components of an efficient draw solution have also been elaborated [26].

Ideal draw solutions such as ammonia and carbon dioxide were identified as very effective, since they have high solubility as gases and are cost effectively separated from water by modest heating distillation [26].

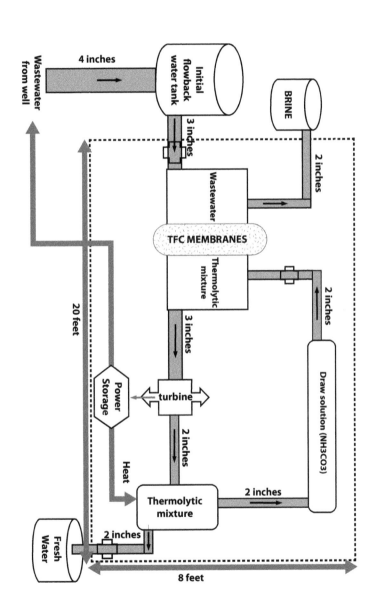

FIGURE 10: Components and flow process outline of the centralized water treatment facility.

7.4.2 FO MEMBRANE SELECTIONS

The major impediment in the widespread use of the forward osmosis application is the availability of a proper membrane system. For a membrane system to be considered economically effective for this purpose, it must be chemically attuned with the chosen draw solution, and both the internal and external concentration polarization of the membrane must have low values. Other factors necessary to consider when choosing a better membrane for the forward osmosis system include; the membrane's working capacity, membrane strength, and membrane configuration. Also considered is the ability to reduce the susceptibility to fouling while increasing both the osmotic pressure and flux. The membrane orientation has been shown to have substantial influence on the performance of the membrane is another major criterion [33]. The thin-film composite (TFC) medium performance membrane was considered for the design since it satisfies the required characteristics for an effective forward osmosis membrane. The membrane is made up of analytical grade Poly-sulfone beads, N, N-dimethyl formamide, 1,3-phenylene-dianmine and 1,3,5-benzenetricarbonyl trichloride [34]. The membrane comprises of a slim active film reinforced with a polymer layer that is very absorbent to water (water flux of 30 L/m²h), with high tendency of rejecting dissolved solutes and is stable up to a pH of 11. The intrinsic permeability (water) of the membrane is 5.81 (L $m^{-2}h^{-1}bar^{-1}$); while the solute permeability coefficient is 0.61 (L $m^{-2}h^{-1}$). The membrane is capable of creating a differential of osmotic pressure up to 25 bars, it has a structural parameter of 370 (μm) and a peak power density of 10 (W/m²) [34]. Based on normal operating conditions, this membrane is estimated to effectively function for five years. The spiral-wound membrane modules are the most effective orientation for membrane design [33]. The application of numerous simultaneous spiral-wound membrane modules increases the efficiency of the water flow treatment, which would enable an optimum recovery rate of 10 bbls/min.

7.5 RESULT AND DISCUSSION: ENGINEERING CONSIDERATIONS

The implementation of water management technologies for hydraulic fracturing water treatment also encompasses sludge solid disposals from the separation processes. The FO process, operating on a hydraulic fracturing flow-back well in the Marcellus Shale area, on average will treat 14, 400 bbls/d (604, 800 gpd) with a chemical composition of 4.3 g/L barium, 219 g/L calcium, 1.3 g/L magnesium, and 3.4 mg/L strontium,. These would produce 40% solids sludge cake, 67,000 lb of barium sludge and 281,815 lb of calcium/strontium/magnesium sludge per day. Pennsylvania Department of Environmental Protection (DEP) has permitting requirements for residual waste and also the cost of moving large quantities of flow back water, hydraulic fracture makeup water, and produced sludge solids, calls for the need to set up a number of dedicated forward osmosis flow back treatment systems to be sited across the area under laid by Marcellus shale formation operations.

Exclusive of the incoming hydraulic fracture flow back water and treated water storage tanks, it is estimated that a 604, 800 gal/d FO process systemswould require a hydro-pneumatic tank and a vessel for the FO system.

7.5.1 GOAL OF THE INTEGRATED SYSTEM

The goal of the proposed design is to provide a portable and effective flow back water treatment system that additionally generates the power necessary to run the system and also eliminate the limitations of currently used techniques. The forward osmosis system designs are flexible, scalable and transportable to facilitate the treatment of flow back water nearby production locations, thus doing away with the hauling costs and environmental contacts related with trucking of the flow back water. A portable shipping container will be modified to house both the Forward Osmosis and Blue Energy portions of the system. An elevated tanker trailer will be used to house the flow back water that will be introduced into the treatment system. This system will be designed to effectively treat 604,800 gallons of flow back water per day. The reclaimed water from the forward osmosis unit would be stored in fracture water tanks located on the well sites which are reused for other fracturing jobs.

7.5.2 FIELD APPLICABILITY OF THE FORWARD OSMOSIS UNIT:

The forward osmosis unit design, (for the approximately 10 bbl/min unit) would have in the low pressure container, a large TFC membrane (1.3 feet (ft) by 3.3 feet spiral-wound components). The FO unit would be efficient in the recovery of about 90% of the flow back water. It has been established that the membranes of this unit are effective in rejecting suspended and dissolved solutes, viruses and bacteria.

A centrifugal transfer pump (0.3 ft by 0.3 ft in dimension), that runs 17 bbl/min would be used to send back the water into the FO unit from the tanks. 36% NH_4CO_3 is circulated into the unit using a 1.8 bbl/min pump. This study has proposed the generation of the needed power from the system. It is imperative to note the forces required for water movement does not come from the pumps, the pressure gradient propels the flow of the water through the membrane. Manpower requirement to operate the 10 bbl/min unit is minimal, for efficiency, a three man schedule can be made where each operator takes an 8 hour shift on allocation.

7.5.3 FORWARD OSMOSIS AND BLUE ENERGY COMBINATION SYSTEM

Initial designs proposed utilized the FO and Blue Energy aspects of the integrated plan as separate entities. These initial designs aimed at increasing the efficiency of the independent systems and then the efficiency of the overall design. After intensive review of the engineering attributes of these previous designs, combining the two systems to work simultaneously to-gether proved to be a more resourceful and cost saving system design. The goal of the Blue Energy in this system is to provide the energy required to pump flow back water from the drilled well and to power the Supervisory Control and Data Acquisition (SCADA) system. The osmotic gradient generated by the TFC-MP membrane and Ammonium bicarbonate draw solution theoretically allows for sufficient energy generation required for the system. This integrated system will decrease the energy costs, increase the effectiveness of the system, and decrease environmental concerns.

7.5.4 TURBINE AND POWER STORAGE CONSIDERATIONS

The power capacity of the system derived through the membrane, when the water from the pressurized section containing the draw solution flows though a water turbine, which in turn generates power. Taking into consid-eration the expected power density of the system, a comparatively small, high efficiency turbine of minimum capacity of 1.2 kW is incorporated in the system. The turbine depends mainly on the flow of the pressurized ther-molytic mixture (mixture of the draw solution and water) and not on depth.

The storage system adapts a capacitor system, with the ability to store power for years. This storage system is connected to a step-up transformer, when required, to raise the power supplied to system equipments. The power storage system also has a minimum power capacity of 1.2 kW.

7.6 SYSTEM OVERVIEW

The proposed design is a novel centralized flow back water treatment facil-ity. This system will be semi-portable to reduce the costs and environment impacts associated with current methods of flow back water treatment of wells in the Marcellus Shale region and other Shale plays. The system will be scaled to treat 604,800 gpd (gallons per day) A Supervisory Control and Data Acquisition (SCADA) system will be utilized to maintain the optimal flow rate of 10 bbl/min of treated water.

The solids separated from the reclaimed water will be contained in a separate tank that can be transported from storage to the market. The treated water will be housed in a separate area. This treated water can be reused in future well fracture operations, sold or can be returned to the environment since it will meet/exceed standards enacted by the Environmental Protection Agency (EPA). Figure 10 shows comprehensively the components of the system and the interconnections that explains the workability of the entire system. The engineering design schematic is presented in figure 11 which illustrates an overview of the most important features of the integrated system design, from the flow back well to the reclaimed water. Forward Osmosis is the most effective flow back water treatment method because it does not require pretreatment of the flow back water and also there is no substantial power requirement in the system, this makes the entire system, environmentally friendly. Forward osmosis has the least carbon signature compared to other flow back water treatment methods that were considered for the flow back water treatment facility which is overly represented in figure 12.

7.6.1 SYSTEM HOUSING SELECTION

A steel portable shipping container will be used to house the entire system consisting of the Forward Osmosis Unit, Blue Energy generation components and SCADA systems. This proposed design will utilize a 40 foot Dual Insulated Dry Goods Shipping Cargo Container with the dimensions 40 ftL×8 ftH×8 ftW. A Dual Insulated Dry Goods Shipping Cargo Container was the ideal housing unit since modified containers are readily available. The estimated price for the 40 ft shipping container and chassis is 7,000 USD including modifications. The Forward Osmosis and Blue Energy systems would be housed within 30-36 ft of the available 40 ft of the Dual Insulated Dry Goods Shipping Cargo Container. The reaming 4-10 ft of available length will be used to house the main components of the SCADA system and other miscellaneous components. The container will need separate insulation considerations for the Forward Osmosis process and for SCADA system. The Forward Osmosis system section of the container must be insulated to prevent the liquids in the system from freezing and to reduce the costs associated with heating required for the separation of the draw solute from the treated water. The SCADA section of the container must be insulated and vented to prevent the computer components from overheating in the warmer periods of operation. During the cooler periods, the computer components will generate enough heat to prevent failure. Anti-Slip mats will be installed within the Dual Insulated dry Goods Shipping Cargo Container to reduce the likelihood that any human operator would slip due to liquids on the floor. Shatterproof fluorescent lights will be installed in both the Forward Osmosis section and SCADA section of the container. The lights will be installed mainly for maintenance operations. This container will be housed upon a trailer designed to house and haul shipping containers by semi trucks. This elevation also allows for housing a container to collect and house the dissolved solutes separated out of the flow back water by the forward osmosis membrane. This collection method will be directly connected to the system to prevent spillage of the dissolved components into the environment. A plastic intermediate bulk container could be used since it is light weight, durable and resists corrosion better than a steel intermediate bulk container.

7.7 IMPACT OF FO INTEGRATED TECHNOLOGY

The extensive use of the FO technology to recover drilling and fracturing wastewater would have some environmental impact as well. This could be a limiting factor, but this is minimal and less impacting than other known methods, considering the fact that the energy necessary to run the system is produced by the system. Another important limiting factor to consider in this design is the chemical composition of the formation, the membrane fouling and dysfunctional part replacement. These problems can be solved during the site specific design of this integrated system to accommodate for these barriers. Ultimately this system is more cost effective than any other conventional methods used presently. If the reserve waste water tank contains 74,000 bbls, and the forward osmosis unit recovers 90% of the flow backwater, then 66,600 bbls off low back water will be recovered/treated and 8,000 bbls would go for recirculation. To reclaim the 66,600 bbls of waste water, the forward osmosis system would require the use

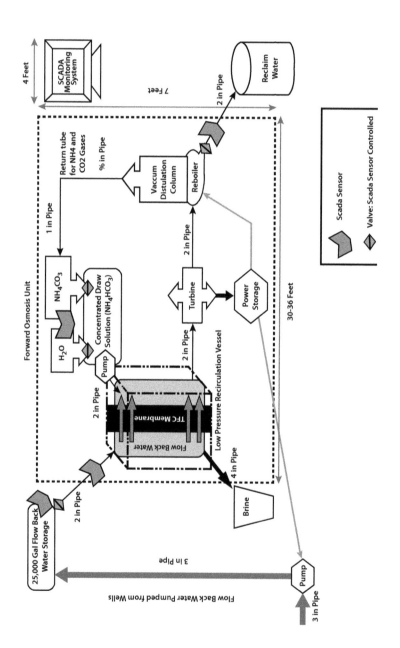

FIGURE 11: Schematic engineering representation of the centralized forward osmosis water treatment facility.

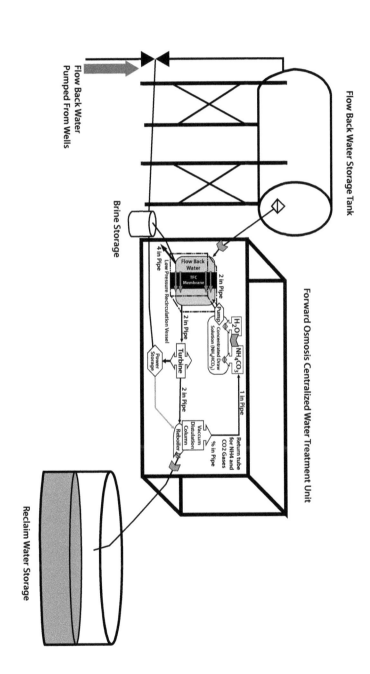

FIGURE 12: Generalized footprint-novel centralized forward osmosis water treatment facility.

of approximately 6000 bbls of 36% NH_4CO_3, leading to a total produced volume of about 72,600 bbls.

Extensive use of the FO process to reclaim drilling and fracturing flow back water in the Marcellus Shale would save approximately 750 million gallons of fresh watery early. Aside this benefit, the FO model also eliminates 66,600 bbls of waste water per horizontal well of hauling related road damages. Approximately 175 truckloads of waste water can be effectively eliminated per reserve pit from conventional practice. Based on the suppositions above, emissions can be drastically reduced with an extensive use of FO/Blue Energy model to regain drilling and fracturing waste fluids, which could save an average of 143,000 bbls/year of diesel. The United States consists of about 25 shale basins where the forward osmosis technology alongside the power generation can be employed. Conversely, the FO system model is not restricted to shale development but also relevant in conventional petroleum exploration areas.

7.8 CONCLUSION

The advent of shale gas development in the quest of meeting the world's energy demands, did not come only with benefits but also some challenges. The natural low permeability shale must be fractured to guarantee higher productivity and the fracturing process involves the use of millions of gallons of water that must be recovered as flow back or produced waste water. Due to pressure from the public and other regulatory agencies, operators in the petroleum industry are expected to comply in the improvement of their water management practices. From analytical results and testing, the FO system is comparatively the easiest, portable and scalable unit with resourcefully and efficiently reclamation capabilities of water waste for valuable reuse. Aside the reduction in the quantity of fresh waterused, the results confirms that forward osmosis can substantially lessen the carbon foot print of exploration and production in the petroleum industry. The combination of FO system and power generation will not only save fresh water resources, but it will provide more water resources that are reusable for other purposes. This integrated system is cost effective and it will improve the public's perception of operator's responsibilities to the environment.

REFERENCES

1. King GE (2010) Thirty Years of Gas Shale Fracturing: What Have We Learned? SPE.
2. Hubbert MK, Willis DGW (1957) Mechanics of hydraulic fracturing. Transaction of the American Institute of Mining, Metallurgical, and Petroleum Engineers Incorporated 210: 153-168.
3. Olawoyin R, Wang JY, Oyewole SA (2012) Environmental Safety Assessment of Drilling Operations in the Marcellus Shale Gas Development, SPE Drilling & Completion, SPE.
4. Engelder T, Lash GG (2008) Marcellus Shale Plays Vast Resource Potential Creating Stir in Appalachia. The American Oil and Gas Reporter.
5. Engelder T (2009) Marcellus 2008: Report card on the breakout year for gas production in the Appalachian Basin. Fort Worth Basin Oil and Gas, August 2009 edition, Abilene, TX, 18-22.
6. Bell CE, Brannon HD, Hughes B (2011) Redesigning Fracturing Fluids for Improving Reliability and Well Performance in Horizontal Tight Gas Shale Applications. SPE.
7. Hurst A, Scott A, Vigorito M (2011) Physical characteristics of sand injectites. Earth-Science Reviews 106: 215-246.
8. Olawoyin RO (2011) Natural Gas Development in the Marcellus Shale. MS thesis, Pennsylvania State University, University Park, Pennsylvania.
9. Davies RJ, Clarke AL (2010) Storage rather than venting after gas hydrate dissociation. Geology 38: 963-966.
10. Zuhlsdorff L, Spie V (2004) Three-dimensional seismic characterization of a venting site reveals compelling indications of natural hydraulic fracturing. Geology 32: 101-104.
11. Arthur JD, Uretsky M, Wilson P (2010) Water Resources and Use for Hydraulic Fracturing in the Marcellus Shale Region. ALL Consulting, Tulsa, OK.
12. Cartwright J, Huuse M, Aplin A (2007) Seal bypass systems. American Association of Petroleum Geologists 91: 1141-1166.
13. Christopherson S (2011) The Economic Consequences of Marcellus Shale Gas Extraction: Key Issues. CaRDI Reports Issue No 14, Community and Regional Development Institute, Cornell University, Ithaca, NY.
14. Kargbo DM, Wilhelm RG, Campbell DJ (2010) Natural gas plays in the Marcellus Shale: challenges and potential opportunities. Environ Sci Technol 44: 5679–5684.
15. Soeder DJ, Kappel WM (2009) Water Resources and Natural Gas Production from the Marcellus Shale. US Geological Survey Fact Sheet 2009–3032, Reston, VA.
16. (2011) US Environmental Protection Agency (USEPA), Plan to Study the Potential Impacts of Hydraulic Fracturing on Drinking Water Resources. Office of Research and Development, Washington DC.
17. Veil J (2010) Water Management Technologies Used by Marcellus Shale Gas Producers. Argonne National Laboratory, Argonne, IL Prepared for the U.S. Department of Energy, National Energy Technology Laboratory.

18. Bourgoyne AT Jr, Millheim KK, Chenevert ME, Young FS Jr (1986) Applied Drilling Engineering (SPE Textbook Series) Society of Petroleum Engineers 2: 85-112.

19. Cipolla CL, Ceramics C (2009) Modeling Production and Evaluating Fracture Performance in Unconventional Gas Reservoirs. Journal of Petroleum Technology, SPE, 84-90.

20. Paktinat J, Pinkgouse JA, Fontaine J (2007) Investigation of Methods to Improve Utica Shale Hydraulic Fracturing in the Appalachian Basin. Society of Petroleum Engineers, SPE.

21. Palisch TT, Vincent M, Handren P (2008) Slickwater Fracturing: Food for Thought. Society of Petroleum Engineers, SPE.

22. (2005) National Oceanic and Atmospheric Administration (NOAA), Annual Summary.

23. Daniel J, Bohm B, Cornue D (2009) Environmental Considerations of Modern Shale Gas Development. Society of Petroleum Engineers, SPE 122931.

24. Gaudlip AW, Paugh LO, Hayes TD (2008) Marcellus Water Management Challenges in Pennsylvania. Paper SPE 119898, presented at the SPE Shale Gas Production Conference, Ft Worth, TX, USA, 16-18 November.

25. Veil JA, Puder MG, Elcock D, Redweik RJ (2004) A White Paper Describing Produced Water from Production of Crude Oil, Natural Gas, and Coal Bed Methane. ANL Report under DOE (NETL) Contract W-31-109-Eng-38.

26. Mc Ginnis R, Mc Cutcheon JR, Elimelech M (2007) A novel ammonia-carbon dioxide osmotic heat engine for power generation. J Membr Sci 305: 13-19.

27. Mc Cutcheon JR, Mc Ginnis RL, Elimelech M (2006) Desalination by ammonia–carbon dioxide forward osmosis: Influence of draw and feed solution concentrations on process performance. J Membr Sci 278: 114–123.

28. Cath TY, Childress AE, Elimelech M (2006) Forward osmosis: Principles, applications, and recent developments. J Membr Sci 281: 70-87.

29. Bamaga OA, Yokochi A, Zabara B, Babaqi AS (2011) Hybrid FO/RO desalination system: Preliminary assessment of osmotic energy recovery and designs of new FO membrane module configurations. Desalination 268: 163-169.

30. Mi B, Elimelech M (2010) Organic Fouling of Forward Osmosis Membranes: Fouling Reversibility and Cleaning without Chemical Reagents. J Membr Sci 348: 337-345.

31. Lee K, Baker R, Lonsdale H (1981) Membrane for power generation by pressure retarded osmosis. J Membr Sci 8: 141-171.

32. Achilli A, Cath TY, Childress AE (2010) Selection of Inorganic-Based Draw Solutions for Forward Osmosis Applications. J Membr Sci 364: 233-241.

33. Wang R, Shi L, Tang CY, Chou S, et al. (2010) Characterization of Novel Forward Osmosis Hollow Fiber Membranes. J Membr Sci 355: 158-167.

34. Yip NY, Tiraferri A, Phillip WA, Schiffman JD, Hoover LA, et al. (2011) Thin-Film Composite Pressure Retarded Osmosis Membranes for Sustainable Power Generation from Salinity Gradients. Environ Sci Technol 45: 4360-4369.

CHAPTER 8

Co-Precipitation of Radium with Barium and Strontium Sulfate and Its Impact on the Fate of Radium during Treatment of Produced Water from Unconventional Gas Extraction

TIEYUAN ZHANG, KELVIN GREGORY, RICHARD W. HAMMACK, AND RADISAV D. VIDIC

8.1 INTRODUCTION

Radium-226/228 is formed by natural decay of uranium-238 and thorium-232 and occurs in natural gas brines brought to the surface following hydraulic fracturing.(1) Because radium is relatively soluble over a wide range of pH and redox conditions, it is the dominant naturally occurring radioactive material (NORM) and an important proxy for radioactivity of waste streams produced during unconventional gas extraction.(2, 3) Radium is a member of alkaline-earth group metals and has properties similar to calcium, strontium, and barium. Oral radium uptake can lead to substitution of calcium in bones and ultimately long-term health risks. Radium-226 activity in Marcellus-Shale-produced water ranges from hundreds to thousands picocuries per liter (pCi/L), with a median of 5350

Reprinted with permission from: Zhang T, Gregory K, Hammack RW, and Vidic RD. Environmental Science and Technology *48,8 (2014). DOI: 10.1021/es405168b. Copyright 2014 American Chemical Society.*

pCi/L.(1) The total radium limit for drinking water and industrial effluents is 5 and 60 pCi/L, respectively.(4)

Radium activity in flowback water from the Marcellus Shale play shows positive correlation with total dissolved solids (TDS) and barium content, despite the differences in reservoir lithologies.(1, 5) This finding is consistent with the fact that the radium/barium ratio is often constant in unconfined aquifers, implying that the radium co-precipitation into barite controls the activity of radium.(6) The high TDS (680–345 000 mg/L)(7) in produced water from Marcellus Shale gas wells is one of the main considerations when choosing a proper radium treatment technology. While there are several treatment options for radium removal, none is as cost-effective in high TDS brines as sulfate precipitation.(8) Despite a very low solubility product for $RaSO_4$ ($K_{sp,RaSO4} = 10^{-10.38}$),(9) it is not likely to observe pure $RaSO_4$ precipitate because of very low radium concentrations in the produced water. However, radium may co-precipitate with other carrier metals.

A distribution equation has been used to describe the co-precipitation of a soluble tracer with a carrier ion. Sulfate-based co-precipitation of radium in a binary system with barium has been examined previously (6, 9-12) and is described by the following distribution equation:

$$\frac{RaSO_4}{MSO_4} = K_d \frac{Ra^{2+}}{M^{2+}}$$

where K_d is the concentration-based effective distribution coefficient, MSO_4 and $RaSO_4$ are relative fractions (or "concentrations") of carrier and radium in solid precipitate, and M^{2+} and Ra^{2+} are equilibrium concentrations in solution. Derivation of the theoretical distribution coefficient with associated thermodynamic parameters is summarized in the Supporting Information and Tables SI-1 and SI-2 of the Supporting Information. Theoretical distribution coefficients of Ra in $BaSO_4$ and $SrSO_4$ in dilute solution are 1.54 and 237, respectively. An increase in the ionic strength (IS) of solution would lead to a decrease in the activity coefficients for Ra^{2+}, Ba^{2+}, and Sr^{2+}, as shown on Figure SI-1 of the Supporting Information. Changes in the activity coefficient ratio of tracer and carrier ion, which is critical when calculating the distribution coefficient in binary systems (see eq 9 of the Supporting Information), are much more pronounced in the case of Sr^{2+} than Ba^{2+} (see Figure SI-1 of the Supporting Information). Consequently, the theoretical distribution coefficient for $Ra–Sr–SO_4$ exhibits more than 50% decline when the IS increased to 3 M, while the decrease in the case of $Ra–Ba–SO_4$ was less than 10% (Figure 1). On the basis of this analysis, it can be expected that an increase in the IS of solution would have a much greater impact on the removal of Ra^{2+} by co-precipitation with $SrSO_4$ than with $BaSO_4$.

Even though the distribution equation has been used to explain the co-precipitation reactions, it has several limitations. First, the presence of electrolytes in solution changes the surface properties (i.e., particle size/ morphology, etch pits, etc.) of the carrier(13) and affects radium removal. Second, the distribution equation assumes that the ions in solution are in equilibrium with the ions throughout the entire solid phase.(14) However, the degree of Ra incorporated into the crystal would be uneven throughout the co-precipitation process if the crystal growth rate is faster than the rate of lattice replacement because the lattice replacement has not reached equilibrium during nucleation and crystal growth. A previous study(15) showed that reduction in the barite precipitation rate significantly increased Ra removal by co-precipitation.

In addition, co-precipitation is a broad term to illustrate the phenomenon where a soluble substance is included in a carrier precipitate, which actually involves three distinct mechanisms: inclusion, occlusion, and adsorption (Figure 2).(15) Inclusion or lattice replacement reaction occurs when a tracer (i.e., Ra^{2+}) occupies a lattice site in the carrier mineral (e.g., barite and celestite), resulting in a crystallographic defect with the tracer in place of the main cation. Occlusion refers to the phenomenon where a tracer is physically trapped inside the crystal during crystal growth, which can be explained by the entrapment of solution or adsorption of tracer during the crystal growth.(15-18) However, occlusion is not likely to play a major role in Ra removal during barite precipitation because of the low moisture content of barite crystal (<3.5%)(16) and because Ra is present in solution at very low levels. Adsorption occurs when the tracer is weakly bound at the surface of the precipitate.(15) As described in the Supporting Information, the distribution equation reflects only the inclusion (lattice replacement) mechanism while neglecting contributions to tracer uptake by adsorption and occlusion. Even though occlusion is a minor mechanism for Ra removal, neglecting adsorption and occlusion during tracer uptake and kinetic effects would inevitably lead to uncertainty in theoretical predictions.(6)

This study focuses on understanding the fundamental mechanisms of Ra co-precipitation in Ba/Sr–SO_4 binary and ternary systems at high saturation levels and different ISs. The mechanisms of inclusion and adsorption for Ra incorporation in the precipitate were distinguished by carefully controlling test conditions, so that the experimentally determined distribution of key species in both Ra–Ba–SO_4 and Ra–Sr–SO_4 co-precipitation experiments can be compared to theoretical predictions and Ra leaching from solids generated during co-precipitation and post-precipitation studies. The impact of precipitation kinetics, activity coefficient ratios, and volumetric mismatch between substituting end-members were analyzed as key factors influencing the fate of radium during co-precipitation with barite and celestite. Additionally, uptake of radium by barite and celestite post-precipitation was compared to the co-precipitation process to understand the relative impact of inclusion, occlusion, and adsorption on the overall radium removal by sulfate precipitation. This study further elucidates fundamental mechanisms influencing the fate of radium during chemical precipitation of divalent cations from produced water (i.e., sulfate precipitation) and associated implications for its reuse for hydraulic fracturing following treatment.

8.2 MATERIALS AND METHODS

Radium-226 source was obtained from Pennsylvania State University and analyzed using a gamma spectrometer(23) with a high-purity germanium detector (Canberra BE 202). Barium chloride dihydrate (99.0% minimum, Mallinckrodt Chemicals), strontium chloride hexahydrate (99.0%, Acros Organics), sodium chloride (99.8%, Fisher Scientific), anhydrous sodium sulfate (100%, granular powder, J.T. Baker), trace-metal-grade nitric acid (65–70%, Fisher), and trace-metal-grade hydrochloric acid (37.3%, Fisher) were American Chemical Society (ACS)-grade. Commercial standards

(Ricca Chemicals and Fisher) were used to calibrate atomic absorption spectrophotometer, and Ultima Gold high-flash-point liquid scintillation counting (LSC) cocktail (PerkinElmer) was used for the liquid scintillation counter. All reagents were tested and found to be free of radium.

The concentration of dissolved Ba and Sr was measured by atomic absorption spectrometry (PerkinElmer model 1000 AAS) with a nitrous oxide–acetylene flame. The filtrate was diluted in a 2% nitric acid and 0.15% KCl solution prior to analysis to limit interferences during metal analysis. Dilution ratios were chosen on the basis of the linear range of this instrument.

Radium-226 activity was analyzed using Packard 2100 LSC through the direct measurement of radium-226.(19) A total of 4 mL of the liquid sample was mixed with 14 mL of Ultima Gold universal LSC cocktail and counted by LSC for 60 min in the specific energy range (170–230 keV) to reject any contribution that is not produced by radium-226. The sample with high IS was corrected by the quench factor, and the ingrowth of radioactivity was compensated by the ingrowth factor.(19, 20) Samples were occasionally calibrated by a gamma spectrometer(21) to ensure accuracy of radium-226 detection, especially at different salinities. Results showed that LSC analysis deviated from gamma spectrometry by less than 7.4%. Activity in both liquid and solid was measured for selected samples to validate the mass balance for radium-226.

Co-precipitation experiments were performed in 50 mL high-density polyethylene (HDPE) tubes. IS was adjusted to 1, 2, or 3 mol/L with concentrated NaCl solution. Radium-226 stock solution was diluted to a target level of 10 000 pCi/L, and the initial Ba^{2+} and Sr^{2+} concentrations were always 5 mmol/L. Different doses of sodium sulfate were added to adjust barium and strontium removal, and pH was not controlled in these experiments. HDPE tubes were placed on a horizontal shaker to promote mixing. Aqueous samples were taken after 24 h of reaction and filtered through 0.45 μm mixed cellulose ester membranes (MF-Millipore, HAWP) prior to analysis for radium-226, barium, and strontium. Because of the relatively slow kinetics of $SrSO_4$ formation,(13) $Ra–Sr–SO_4$ solutions were sampled after 5, 24, and 48 h.

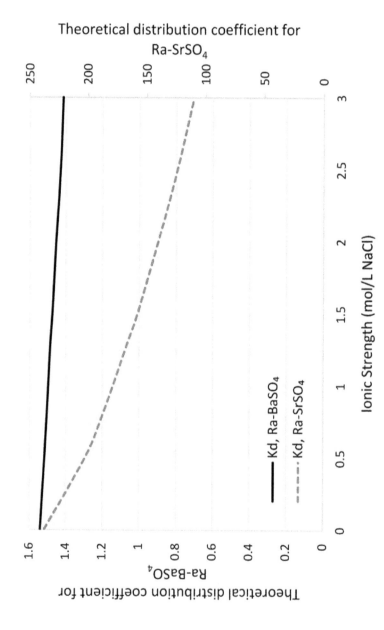

FIGURE 1: Theoretical distribution coefficient (Kd,Ra–MSO$_4$) for radium in BaSO$_4$ and SrSO$_4$ as a function of the IS based on eq 9 of the Supporting Information.

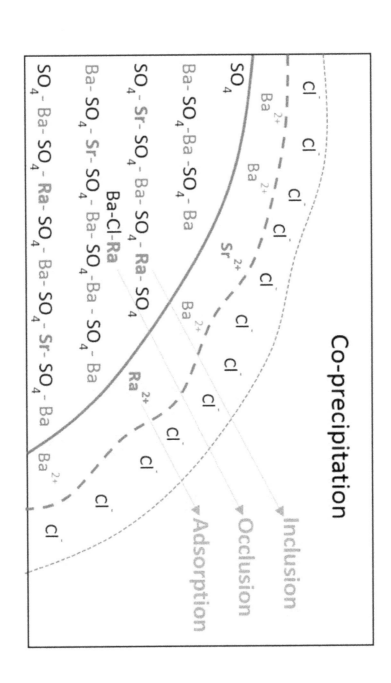

FIGURE 2: Three mechanisms (inclusion, occlusion, and adsorption) of radium co-precipitation in binary solution with Ba–SO$_4$

TABLE 1: Experimental Conditions for Ra Removal in Binary and Ternary Systems(32)[a]

system	initial concentration				IS (mol)	ion activity			SI	
	NaCl (mol/L)	Ba (mmol/L)	Sr (mmol/L)	SO$_4$ (mmol/L)		Ba (mmol/L)	Sr (mmol/L)	SO$_4$ (mmol/L)	SIBaSO$_4$	SISrSO$_4$
Ba–Ra–SO4 binary	0	5	0	0.5	0.0165	3.155		0.173	3.71	
	0	1.25	0.01875	2.943		0.483				
	0	5	0.03	2.146		1.9				
	5	0	0.5	3.0165	1.276			2.22		
	0	1.25	3.01875	1.277		0.349				
	0	5	3.03	1.285		0.139				
Sr–Ra–SO4 binary	0	0	5	1.25	0.01875		4.09	0.537		0.85
	5	5	0.03		2.323	2.143	4.58	1.33		
	5	10	0.045		1.773	4.164	0.02	1.51		
	0	5	5	3.03		2.246	2.62	0.42	0.13	
	5	10	3.045		2.234	0.273	3.22	0.72		
	5	20	3.075		2.21	0.542	3.01	1.09		
	5	50	3.165		2.142	1.324	0.137			
Ba–Sr–Ra–SO4 ternary	0	5	5	1.25	0.03375	2.674	2.767	0.339	3.92	0.6
	5	5	0.045	2.166	2.417	1.453	4.47	1.18		
	5	10	0.06	1.702	2.016	3.064	4.69	1.43		
	3	5	1.25	3.03375	1.279	2.255	0.035	2.62	-0.47	
	5	5	3.045	1.286	2.246	1.382	3.22	0.13		
	5	10	3.06	1.296	2.233	0.275	3.52	0.43		

[a]$SI = log(IAP/KSP)$, where IAP is the ion activity product and KSP is the solubility product. IS was adjusted with NaCl. Activity coefficients were calculated using Pitzer equation.

Radium removal by barite/celestite post-precipitation was studied by adding a specific amount of preformed solids (barite and/or celestite) into 10 000 pCi/L radium-226 solution. Barite and celestite were prepared from the solution composition that is identical to that used in co-precipitation experiments to ensure identical particle morphology and size. After 24 h of moderate shaking, aqueous samples were removed and filtered through a 0.45 μm membrane prior to radium-226 analysis.

Experiments were performed to examine the equilibrium and kinetics of Ra–Ba–SO$_4$ and Ra–Sr–SO$_4$ formation alone and in combination using the initial conditions listed in Table 1. The binary and ternary systems were studied at high IS to simulate radium removal from brines generated by unconventional gas extraction. The distribution coefficient was calculated for each system and compared to theoretical values. Both kinetics and equilibrium studies were conducted to provide a fundamental understanding of the fate of radium during chemical precipitation, employed to remove divalent cations from natural gas brines (i.e., sulfate precipitation) and facilitate its reuse for hydraulic fracturing.

8.3 RESULTS AND DISCUSSION

8.3.1 IMPACT OF IS ON RA REMOVAL BY CO-PRECIPITATION
IN BINARY SYSTEMS

The radium removal and experimental distribution coefficient (K_d') for Ra–Ba–SO$_4$ co-precipitation at different ISs is shown in Figure 3a. Radium co-precipitation in dilute solutions (i.e., IS of about 0.02 was due to the addition of BaCl$_2$ and Na$_2$SO$_4$ only) was proportional to barium removal, which can be described by the distribution law. The decrease of K_d' with the increase in Ba removal is expected because the inclusion of Ra into BaSO$_4$ during the initial stages of BaSO$_4$ precipitation decreases the Ra concentration in solution, resulting in a much lower Ra concentration to co-precipitate with subsequent BaSO$_4$. The experimental distribution coefficient (K_d' = 1.07–1.54) was always below the theoretical value (K_d = 1.54) in dilute solutions, which can be attributed to the fast barite crystal growth at high supersaturation levels used in these experiments [saturation index $(SI) = 3.7$–4.6]. Under these conditions, barite precipitation was completed within just 10 min, which adversely impacts radium removal because inclusion and occlusion processes only occur during nucleation and crystal growth of barite. Rosenberg et al.(22) reported that experimental K_d' can be as high as 3 when precipitation kinetics is controlled by continuously adjusting the concentration of reactants in the solution.

The dependence of the distribution coefficient upon IS (Figure 1) suggests only a slight decrease in K_d for Ra–Ba–SO_4 with an increase in IS. However, experimental results show that radium co-precipitation was enhanced in the presence of electrolytes, with experimental K_d' increasing to 3.17 at IS of 1.02 M and 7.49 at IS of 3.03 M for barium removal of 10% (Figure 3a). Such high values of the distribution coefficient cannot be explained by thermodynamics of lattice replacement reactions. It has been reported that the solubility of $BaSO_4$ increases with IS,(23, 24) which would lead to a decrease in the equilibrium constant, as shown by eq 6 of the Supporting Information. However, the solubility of $RaSO_4$ would also increase with IS, which would offset the increase in $BaSO_4$ solubility. Hence, a change in the thermodynamic driving force at high salinity is an unlikely reason for enhanced radium removal.

There are several explanations for the increase in radium removal with an increase in IS. First, the activities of electrolytes decrease with an increase in IS (Table 1), which reduces supersaturation. Because nucleation of $BaSO_4$ follows the homogeneous nucleation theory with diffusion-controlled crystal growth,(25-27) a decrease in supersaturation leads to a sharp decrease in the nucleation rate(28) and a decrease in the crystal growth rate.(26) This reduction in the rate of precipitation would enhance incorporation of radium into $BaSO_4$ because it would allow more time for lattice replacement reactions during the crystal growth. In addition, the increase in IS would decrease the crystal–solution interfacial tension,(28) increase etch density,(29) and compress the electric double layer,(30) which increases the probability of the Ra^{2+} reaction with the $BaSO_4$ lattice.

A high distribution coefficient for Ra–$SrSO_4$ co-precipitation (Figure 1) is attributed to large differences in solubility products of $RaSO_4$ ($K_{sp,RaSO4}$ = 10–10.38) and $SrSO_4$ ($K_{sp,SrSO4}$ = 10–6.63). However, the possibility of the inclusion reaction decreases when the volumetric mismatch between the two end members (i.e., $RaSO_4$ and $SrSO_4$) is large (see Table SI-2 of the Supporting Information),(31) which would significantly depress radium incorporation into $SrSO_4$ precipitate. The mismatch phenomenon can be quantified by the Margules parameter (W), as described in Table SI-2 of the Supporting Information. The Margules-corrected distribution coefficient for Ra–$SrSO_4$ (K_d = 237) is very large compared to that for Ra–$BaSO_4$ (K_d = 1.54), which implies that $SrSO_4$ should have stronger affinity for radium. Experimental results (Figure 4a) show that radium removal in dilute solutions is always around 80%, regardless of Sr removal. Consequently, the experimental distribution coefficient for Ra–$SrSO_4$ varies from 43 to below 1 (Figure 4b) and is much lower than the theoretical value.

A significant decrease in activity coefficient ratios of ($\gamma_{Ra2+}/\gamma_{M2+}$) at elevated IS (see Figure SI-1 of the Supporting Information) would reduce the theoretical distribution coefficient. The theoretical distribution coefficient for Ra–$SrSO_4$ at IS = 3 M of 110 is still very large compared to Ra–$BaSO_4$ (Figure 1). Experimental results show that radium removal is greater than 75%, as long as Sr removal is greater than 8% (Figure 4a). The discrepancy of K_d and K_d' is attributed to the kinetic limit for Ra inclusion into $SrSO_4$ and underestimation of incompatibility (volumetric mismatch) of Ra–SO_4 in Sr–SO_4 lattice, which limits Ra removal at relatively short reaction times (<48 h) and exacerbates the competition of Ra with other cations for the lattice replacement reaction.

8.3.2 RA REMOVAL BY CO-PRECIPITATION IN A TERNARY SYSTEM

In actual flowback and produced water from unconventional gas extraction, both barium and strontium are present at concentrations that are of the same order of magnitude.(7) Synthetic solutions used for the study of the ternary system contained 10 000 pCi/L of radium and 5 mM each of barium and strontium, and the IS was adjusted using sodium chloride. Sulfate dosage between 1.25 and 10 mM was added to control Ra–Ba–Sr–SO_4 precipitation, and radium removal was compared to barium removal as a function of IS (Table 1).

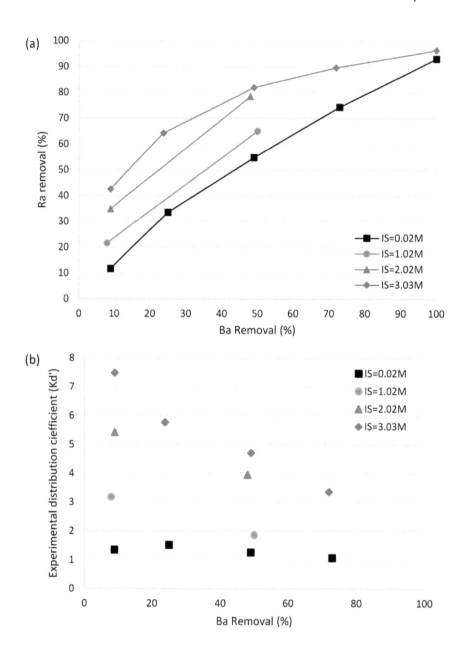

FIGURE 3: Radium co-precipitation with BaSO$_4$ as a function of barium removal at different ISs adjusted with NaCl: (a) radium removal and (b) experimental distribution coefficient, at pH 7 and 5 mM Ba^{2+}initial. Ba removal was adjusted with sulfate addition.

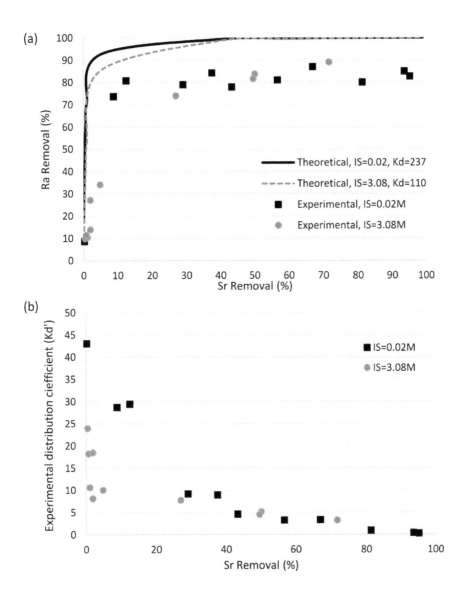

FIGURE 4: Radium co-precipitation with $SrSO_4$ as a function of strontium removal at different ISs adjusted with NaCl: (a) radium removal and (b) experimental distribution coefficient, at pH 7 and 5 mM Sr^{2+}initial. Sr removal is adjusted with sulfate addition (1.25–10 mM for dilute system and 5–50 mM for IS ≈ 3 M).

Because both $BaSO_4$ and $SrSO_4$ are good radium carriers, overall radium removal in the ternary system was expected to be enhanced by the synergy of the two co-precipitation processes. However, kinetics of $BaSO_4$ precipitation was much faster than that of $SrSO_4$ under the experimental conditions used in this study because the SI for $BaSO_4$ (from 2.6 to 4.7) was much higher than that for $SrSO_4$ (from −0.47 to 1.43). A previous study[32] showed that the kinetics of $BaSO_4$ precipitation under similar conditions was much faster than $SrSO_4$ (i.e., $BaSO_4$ precipitation was completed within 30 min, while it took several days for $SrSO_4$ to reach equilibrium). It is expected that faster $BaSO_4$ precipitation is likely to control radium removal by inclusion in the precipitate.

As shown in Figure 5, the dependence of radium removal upon barium removal in the Ra–Ba–Sr–SO_4 ternary system follows that for the Ra–Ba–SO_4 binary system. A slight decrease in Ra removal observed in the ternary system can be attributed to the presence of Sr that competes with radium for co-precipitation with $BaSO_4$.[33]

TABLE 2: Radium, Barium, and Strontium Dissolution from Solids Generated in Binary and Ternary Co-precipitation Systems after 24 h at pH 0.5[a]

sample	initial concentration (mmol/L)			solids concentration (mg/L)	fraction dissolved (%)		
	Ba	Sr	SO_4		Ba	Sr	Ra
Ba–Ra–SO_4 binary	5		5	1167	3.0		5.4
Sr–Ra–SO_4 binary		5	5	918		47.0	73.3
Ba–Sr–Ra–SO4 ternary	5	5	10	2085	3.0	51.0	6.7

[a]The initial Ra concentration in all tests was 104 pCi/L.

To verify that $BaSO_4$ is the main Ra carrier in the Ba–Sr–SO_4 system, precipitates created in binary and ternary systems were collected on a 0.45 μm filter membrane and added into 50 mL of 5 mM barium and strontium solution to suppress readsorption of radium on the remaining solids. Hydrochloric acid was then added to adjust pH to 0.5 and dissolve $SrSO_4$. After that, aqua regia was added to dissolve any remaining solids and radium mass balance closure above 80% was required to accept the results from these tests. Dissolution of Ra, Ba, and Sr from the solid phase at pH 0.5 is summarized in Table 2.

The results for sample A obtained using the solids precipitated in the Ra–Ba–SO_4 binary system show that very little radium was released into solution (5.4%) at pH 0.5 when $BaSO_4$ was the only radium carrier. This is expected because there was minimal (3.0%) $BaSO_4$ dissolution at pH 0.5. The test with sample B that was obtained using the solids precipitated in the Ra–Sr–SO_4 binary system showed that strontium dissolution was significant at pH 0.5 (47.0%) and that a large fraction of radium was released into the solution (73.3%) under these conditions. A higher percentage of radium released to solution compared to strontium indicates that radium is not tightly bound in the $SrSO_4$ lattice, which can be explained by the large volumetric mismatch between the two (see Table SI-2 of the Supporting Information). In sample C collected from the Ra–Ba–Sr–SO_4 ternary system, only 6.7% of radium was released to the solution after 24 h at pH 0.5, while the fractions of strontium (51.0%) and barium (3.0%) released to the solution were similar to those observed in the case of binary systems. Very low radium release from solids collected in both Ra–Ba–SO_4 and Ra–Ba–Sr–SO_4 systems confirmed that radium is mainly bound to $BaSO_4$ solids during Ra–Ba–Sr–SO_4 co-precipitation.

8.3.3 CO-PRECIPITATION VERSUS POST-PRECIPITATION FOR RADIUM REMOVAL

Co-precipitation is defined as simultaneous removal of both tracer and carrier from an aqueous solution and is due to inclusion (lattice replacement), occlusion, and adsorption reactions (Figure 2). The term post-precipitation refers to tracer removal by a previously formed carrier precipitate when only lattice replacement and adsorption are feasible removal mechanisms. Removal of Ra by preformed barite and celestite may be an important mechanism for Ra sequestration in a treatment process that uses solids recycling to enhance the precipitation kinetics in the reactor and was evaluated in this study using the experimental conditions outlined in Table 3.

TABLE 3: Radium Post-precipitation Removal by Preformed Barite and Celestite[a]

adsorbent	solid concentration (g/L)	solution composition	Ra removal after 24 h (%)	Ra desorption ratio[b] (%)	Ra desorption (pCi/L)[c]
barite	0.2	DI water	84.3	36.2	3052
	0.5	DI water	84.0	19.3	1621
	1	DI water	87.2	11.6	1012
	1	5 mM Ba	32.0	24.8	794
	1	5 mM Ba; 5 mM Sr	29.5	26.4	779
	1	3 M NaCl	94.8	4.1	389
	5	5 mM Ba	66.2	15.6	1033
	10	5 mM Ba	81.9	14.1	1155
celestite	1	DI water	85.9		
	1	5 mM Sr	52.7		
	1	5 mM Ba	69.8		

[a]All samples were equilibrated for 24 h. The initial Ra concentration in all tests was 104 pCi/L. [b]Ra desorption ratio denotes the desorbed amount as a percentage of total Ra present in the carrier. [c]Ra desorption denotes the total activity of Ra desorbed from the carrier.

The first set of experiments revealed that radium post-precipitation removal by barite did not change much even as barite concentration varied from 0.2 to 1 g/L. In addition, when radium-enriched barite was returned into a fresh radium solution [10 000 pCi/L in deionized (DI) water], radium removal was the same as for a freshly prepared barite (Table 4). Such behavior can be explained by the fact that the impurities (i.e., radium) in the $BaSO_4$ lattice are always negligible ($<2.6 \times 10^{-8}$ g of Ra/g of barite), even after 5 cycles of barite reuse, which makes fresh and reused barite identical in terms of their ability to remove radium.

To identify the extent of radium adsorption on preformed solids in comparison to inclusion, desorption studies were performed at pH 0.5 for 24 h. The desorption ratio is defined as the fraction of total radium in the solids that is released into the solution. Table 3 shows that most radium was strongly bound to the barite lattice under experimental conditions evaluated in this study. The desorption ratio decreased with increasing barite dose, suggesting that adsorption is a less significant radium removal mechanism during post-precipitation compared to inclusion.

Ra post-precipitation removal by preformed barite is strongly suppressed in the presence of Ba in solution (Table 3) because of the competition for inclusion into the barite matrix. The adverse impact of Sr in solution is not as substantial because of significant volumetric mismatch between $BaSO_4$ and $SrSO_4$. IS has a similar impact on radium incorporation into barite in the case of post-precipitation (Table 3) as it did in the case of co-precipitation (Figure 3), as demonstrated by an increase in radium removal with an increase in IS.

TABLE 4: Post-precipitation of Radium in Recycled Barite in DI Water[a]

adsorbent	solid amount (g/L)	initial Ra concentration in barite (pCi of Ra/g of barite)	solution composition (pCi/L Ra)	Ra removal (%)
barite	1	0	10000	84.30
	1	8430		87.47
	1	17177		84.87
	1	25664		85.07

[a]All samples were measured after 24 h.

Radium removal by the preformed celestite was strongly depressed in the presence of competition ions (i.e., strontium or barium). This phenomenon was expected because the effective solid–solution interface area for inclusion reactions is limited in the absence of the crystal growth phase during post-precipitation uptake of radium. However, the decrease in radium removal in the presence of competing ions is less pronounced in comparison to $BaSO_4$ post-precipitation, which is expected because of the very high theoretical distribution coefficient for Ra–$SrSO_4$ and much lower solubility of $SrSO_4$.(34) Desorption of Ra–$SrSO_4$ was not evaluated because celestite is largely dissolved at pH 0.5.

8.3.4 IMPLICATIONS FOR FLOWBACK/PRODUCED WATER TREATMENT BY SULFATE PRECIPITATION

Flowback/produced water generated from Marcellus Shale gas extraction was initially treated in municipal wastewater treatment facilities that are generally not capable of removing TDS, and high conductivities were reported in the Monongahela River basin(35) as a result of this practice. The Pennsylvania Department of Environmental Protection then issued a request in mid-2011 to exclude municipal treatment facilities from this practice and industry complied.(36) Centralized waste treatment (CWT) facilities play a major role in treatment of Marcellus Shale wastewater prior to disposal or reuse in subsequent hydrofracturing operations.(36, 37) The volume of unconventional gas wastewater treated in these facilities increased from 644.4 million liters in 2008 to 1752.8 million liters in 2010.(37) Sulfate precipitation is a common practice in CWT facilities for barium, strontium, and radium removal.

On the basis of the behavior of the Ra–Ba–Sr–SO_4 ternary system at high IS documented in this study, it can be concluded that Ra inclusion in $BaSO_4$ is likely the primary mechanism for its removal in CWT facilities that employ sulfate precipitation. The experimental distribution coefficient for Ra in $BaSO_4$ ranges from 1.07 to 1.54 for dilute solution and from 1.86 to 7.49 at IS $\approx$ 3 M, suggesting that Ra removal in CWT facilities will be higher than Ba removal. This study also suggests that it would be beneficial to recycle barite solids in the treatment process to enhance Ra removal because recycled barite (i.e., Ra-enriched barite) showed very similar Ra removal compared to freshly prepared barite (Table 4). Once radium is incorporated into the barite lattice, it is unlikely to desorb, even at very low pH (e.g., pH 0.5).

A recent study on the impact of shale gas wastewater disposal on water quality in western Pennsylvania revealed elevated levels of radium in sediments at the point of discharge from a CWT facility.(38) Because the CWT evaluated in that study employed sulfate precipitation to achieve over 90% Ba removal, it is expected that the Ra concentration in the effluent would be about 3 orders of magnitude lower than that in the raw wastewater. A continuous low level flux of Ra into the receiving stream would lead to an increase in the Ra content of the sediments downstream of the discharge point.(38) It is also possible that some of the Ra discharge into the receiving stream would be in the form of barite solids containing co-precipitated Ra that were not captured in the CWT. A high density of barite (4.5 g/cm^3) would lead to a fairly limited transport of insoluble barite downstream of the CWT and contribute to TENORM buildup in the river sediments.

Assuming an average initial Ra and Ba concentration in flowback water treated at a CWT facility of 3000 pCi/L(38) and 5 mmol/L,(7) respectively, the estimated level of Ra activity in precipitates would range from 2571 to 18087 pCi/g of $BaSO_4$, depending upon the Ba removal and distribution coefficient (Figure 6). In comparison to TENORM limits for municipal waste landfills, which range from 5 to 50 pCi/g depending upon state regulations (http://www.tenorm.com/regs2.htm), Ra levels in the solids produced in these CWT facilities far exceed these limits. Municipal waste landfills are the main disposal alternative for this solid waste as long as they do not exceed allowed source term loading (ASTL) for TENORM on an annual basis.(39) Sustainable management of solid radioactive waste produced in these treatment facilities may require alternative management strategies. One potential approach to avoid the creation of Ra-enriched solid waste is to use carbonate precipitation for Ra removal because, unlike barite, carbonate solids generated by the treatment plant could be dissolved in mildly acidic solution and disposed by deep well injection.(40) Another alternative is to reuse the Ra-enriched barite generated at CWTs used as a weighting agent in drilling mud that is typically added to maintain the integrity of the wellbore.(41)

It is also important to note that municipal landfills only use γ radiation to monitor the TENORM in the incoming waste,(37) which provides only a rough estimate of Ra-226 activity because it is influenced by the composition of other radionuclides in the waste stream. A comprehensive analysis of the fate of Ra disposed in municipal solid waste landfills is needed to properly assess radiation exposure risks.(42) These risks will be associated with the emission of volatile progenies (i.e., Rn) because the results of this study suggest that Ra will not leach out in a relatively mildly acidic environment of municipal solid waste landfills(43) once it is sequestered in barite solids.

REFERENCES

1. Rowan, E., Engle, M., Kirby, C., and Kraemer, T. Radium Content of Oil-and Gas-Field Produced Waters in the Northern Appalachian Basin (USA)—Summary and Discussion of Data; U.S. Geological Survey Scientific Investigations Report 2011-5135; U.S. Geological Survey: Reston, VA, 2011.

2. Vidic, R. D.; Brantley, S. L.; Vandenbossche, J. M.; Yoxtheimer, D.; Abad, J. D.Impact of shale gas development on regional water quality Science 2013, 340, 6134

3. Grundl, T.; Cape, M.Geochemical factors controlling radium activity in a sandstone aquifer Ground Water 2006, 44 (4) 518– 527,

4. U.S. Nuclear Regulatory Commission. Standards for Protection against Radiation—Appendix B—Radionuclide Table-Radium-226; U.S. Nuclear Regulatory Commission: Rockville, MD, 2013; 10 CFR, 20.

5. Gregory, K. B.; Vidic, R. D.; Dzombak, D. A.Water management challenges associated with the production of shale gas by hydraulic fracturing Elements 2011, 7 (3) 181– 186

6. Gordon, L.; Rowley, K.Coprecipitation of radium with barium sulfate Anal. Chem. 1957, 29 (1) 34– 37

7. Barbot, E.; Vidic, N. S.; Gregory, K. B.; Vidic, R. D.Spatial and temporal correlation of water quality parameters of produced waters from devonian-age shale following hydraulic fracturing Environ. Sci. Technol. 2013, 47 (6) 2562– 2569

8. Fakhru'l-Razi, A.; Pendashteh, A.; Abdullah, L. C.; Biak, D. R. A.; Madaeni, S. S.; Abidin, Z. Z.Review of technologies for oil and gas produced water treatment J. Hazard. Mater. 2009, 170 (2) 530– 551,

9. Langmuir, D.; Riese, A. C.The thermodynamic properties of radium Geochim. Cosmochim. Acta 1985, 49 (7) 1593– 1601

10. Li, M. Removal of divalent cations from marcellus shale flowback water through chemical precipitation. Master's Dissertation, University of Pittsburgh, Pittsburgh, PA, 2011.

11. Prieto, M.Thermodynamics of solid solution–aqueous solution systems Rev. Mineral. Geochem. 2009, 70 (1) 47– 85

12. Doerner, H. A.; Hoskins, W. M.Co-precipitation of radium and barium sulfates J. Am. Chem. Soc. 1925, 47 (3) 662– 675

13. Risthaus, P.; Bosbach, D.; Becker, U.; Putnis, A.Barite scale formation and dissolution at high ionic strength studied with atomic force microscopy Colloids Surf., A 2001, 191 (3) 201– 214

14. Gordon, L.; Reimer, C. C.; Burtt, B. P.Distribution of strontium within barium sulfate precipitated from homogeneous solution Anal. Chem. 1954, 26 (5) 842– 846

15. Harvey, D. Modern Analytical Chemistry; McGraw-Hill: New York, 2000.

16. Nichols, M. L.; Smith, E. C.Coprecipitation with barium sulfate J. Phys. Chem. 1941, 45 (3) 411– 421

17. Fischer, R. B.; Rhinehammer, R. B.Rapid precipitation of barium sulfate Anal. Chem. 1953, 25 (10) 1544– 1548

18. Schneider, F.; Rieman, W., IIIThe mechanism of coprecipitation of anions by barium sulfate J. Am. Chem. Soc. 1937, 59 (2) 354– 357

19. Blackburn, R.; Al-Masri, M. S.Determination of radium-226 in aqueous samples using liquid scintillation counting Analyst 1992, 117 (12) 1949– 1951

20. Applications of Liquid Scintillation Counting; Horrocks, D., Ed.; Academic Press: Waltham, MA, 1974.

21. Johnston, A.; Martin, P.Rapid analysis of 226Ra in waters by gamma-ray spectrometry Appl. Radiat. Isot. 1997, 48 (5) 631– 639,

22. Rosenberg, Y. O.; Metz, V.; Ganor, J.Co-precipitation of radium in high ionic strength systems: 1. Thermodynamic properties of the Na–Ra–Cl–SO4–H2O system—Estimating Pitzer parameters for RaCl2 Geochim. Cosmochim. Acta 2011, 75 (19) 5389– 5402

23. Bokern, D. G.; Hunter, K. A.; McGrath, K. M.Charged barite–aqueous solution interface: Surface potential and atomically resolved visualization Langmuir 2003, 19 (24) 10019– 10027

24. Meissner, H. P.; Tester, J. W.Activity coefficients of strong electrolytes in aqueous solutions Ind. Eng. Chem. Process Des. Dev. 1972, 11 (1) 128– 133

25. Fernandez-Diaz, L.; Putnis, A.; Cumberbatch, J.Barite nucleation kinetics and the effect of additives Eur. J. Mineral. 1990, 2 (4) 495– 501[CAS]

26. Nielsen, A. E.; Toft, J. M.Electrolyte crystal growth kinetics J. Cryst. Growth 1984, 67 (2) 278– 288

27. He, S.; Oddo, J. E.; Tomson, M. B.The nucleation kinetics of barium sulfate in NaCl solutions up to 6 m and 90 °C J. Colloid Interface Sci. 1995, 174 (2) 319– 326

28. Anderson, G. M.; Crerar, D. A. Thermodynamics of Geochemistry: The Equilibrium Model; Oxford University Press: New York, 1993; Vol. 588.

29. Risthaus, P.; Bosbach, D.; Becker, U.; Putnis, A.Barite scale formation and dissolution at high ionic strength studied with atomic force microscopy Colloids Surf., A 2001, 191 (3) 201– 214

30. Hang, J. Z.; Zhang, Y. F.; Shi, L. Y.; Feng, X.Electrokinetic properties of barite nanoparticles suspensions in different electrolyte media J. Mater. Sci. 2007, 42 (23) 9611– 9616

31. Zhu, C.Coprecipitation in the barite isostructural family: 1. Binary mixing properties Geochim. Cosmochim. Acta 2004, 68 (16) 3327– 3337

32. He, C.; Li, M.; Liu, W.; Barbot, E.; Vidic, R. D.Kinetics and equilibrium of barium and strontium sulfate formation in Marcellus Shale flowback water J. Environ. Eng. 2014, 10.1061/(ASCE)EE.1943-7870.0000807

33. Ceccarello, S.; Black, S.; Read, D.; Hodson, M. E.Industrial radioactive barite scale: Suppression of radium uptake by introduction of competing ions Miner. Eng. 2004, 17 (2) 323– 330

34. Brower, E.Synthesis of barite, celestite and barium–strontium sulfate solid solution crystals Geochim. Cosmochim. Acta 1973, 37 (1) 155– 158

35. Kargbo, D. M.; Wilhelm, R. G.; Campbell, D. J.Natural gas plays in the Marcellus Shale: Challenges and potential opportunities Environ. Sci. Technol. 2010, 44 (15) 5679– 5684

36. Maloney, K. O.; Yoxtheimer, D. A.Production and disposal of waste materials from gas and oil extraction from the Marcellus Shale play in Pennsylvania Environ. Pract. 2012, 14 (4) 278– 287

37. Lutz, B. D.; Lewis, A. N.; Doyle, M. W.Generation, transport, and disposal of wastewater associated with Marcellus Shale gas development Water Resour. Res. 2013, 49 (2) 647– 656

38. Warner, N. R.; Christie, C. A.; Jackson, R. B.; Vengosh, A.Impacts of shale gas wastewater disposal on water quality in western Pennsylvania Environ. Sci. Technol. 2013, 47 (20) 11849– 11857

39. Pennsylvania Department of Environmental Protection. Final Guidance Document on Radioactivity Monitoring at Solid Waste Processing and Disposal Facilities; Pennsylvania Department of Environmental Protection: Harrisburg, PA, 2004; http://www.elibrary.dep.state.pa.us/dsweb/Get/Document-48337/250-3100-001.pdf.

40. Research Partnership to Secure Energy for America (RPSEA). Produced Water Treatment for Water Recovery and Salt Production; RPSEA: Sugar Land, TX, 2012; 08122-36-Final Report.

41. Oakley, D.; Cullum, D. Advanced Technology Makes New Use of Age-Old Drilling Fluid Agent; Drilling Contractor: Houston, TX, May/June 2007; http://www.drillingcontractor.org/dcpi/dc-mayjune07/DC_May07_MISWACO.pdf.

42. Smith, K. P.; Arnish, J. J.; Williams, G. P.; Blunt, D. L.Assessment of the disposal of radioactive petroleum industry waste in nonhazardous landfills using risk-based modeling Environ. Sci. Technol. 2003, 37 (10) 2060– 2066

43. United States Environmental Protection Agency (U.S. EPA). Method 1311. Toxicity Characteristic Leaching Procedure; U.S. EPA: Washington, D.C., 1992; http://www.epa.gov/osw/hazard/testmethods/sw846/pdfs/1311.pdf.

Figures 5 and 6, along with several supplemental files, are not available in this version of the article. To view this additional information, please use the citation on the first page of this chapter.

PART IV

FRACKING WASTEWATER REGULATIONS

Regulation of Water Pollution from Hydraulic Fracturing in Horizontally-Drilled Wells in the Marcellus Shale Region, USA

HEATHER HATZENBUHLER AND TERENCE J. CENTNER

9.1 INTRODUCTION

In the last four years, horizontal drilling using many fractures along a horizontal wellbore has been used commercially to access the deepest shale gas (over 1800 m below the surface) in the United States [1,2]. Horizontal drilling employs turning a downward-plodding drill bit to continue drilling within a layer underneath the ground. Accompanying horizontal drilling is hydraulic fracturing, a well-stimulation technique that maximizes extraction of oil and natural gas in unconventional reservoirs such as shale, coalbeds and tight sands. Hydraulic fracturing involves injecting specially engineered fluids consisting of chemicals and granular material into the wells at incredible pressure to break up the fuel stores and stimulate the flow of natural gas or oil to the surface [1]. Once the well has been fractured, the pressure forces out some of the injection fluids containing chemicals, brines, metals, radionuclides and hydrocarbons [3]. For some wells, the toxic flowback fluids are removed and later injected into class II injection wells [4]. In other situations, the fluids are recycled or are transported to local wastewater treatment facilities. As a result of horizontal drilling, there has been a significant increase in the natural gas supply and a reduction in wholesale spot price of natural gas by nearly 50% [5].

The risks associated with all aspects of fracturing have been looked at from a variety of perspectives, but most concerns revolve around the use of water resources and their potential contamination [6]. Other risks are associated with surface spills [7,8]. The United States Environmental Protection Agency (EPA) has been investigating drinking water contamination and is expected to complete an extensive study on all aspects of hydraulic fracturing in 2014 [9]. A conclusion that may be drawn from a review of recent scientific studies and incidences is that horizontal drilling accompanied by hydraulic fracturing poses threats to local environmental conditions and the health and safety of persons using land, water, and air resources.

9.2 FEDERAL AND REGIONAL POLICIES

Several federal and regional policies have been adopted to oversee potential risks related to hydraulic fracturing. However, amendments to the federal laws have limited the federal government's oversight of activities accompanying the development of shale gas resources. An overview of relevant legislation, summarized in Table 1, enumerates the role EPA and other agencies could play in minimizing negative impacts of natural gas production.

In 1972, the Clean Water Act (CWA) delineated the basic structure for regulating discharges of pollutants into waters and for establishing quality standards for surface waters under the authority of EPA [10]. Under the CWA's National Pollutant Discharge Elimination System program, stormwater permits were required for sediment runoff from construction sites and discharges of pollutants into surface waters [11]. The permitting system requires adoption of technology-based and water quality-based effluent limits [11,12]. Fracturing activities that inject liquid into the ground or store waters in temporary pits without any discharge are not regulated under the CWA. Thus, there is no federal oversight of fracturing activities until there is proof of fracturing contaminants in surface waters [13].

TABLE 1: Summary of federal and regional legislation.

Legislation	Authority/Jurisdiction	Potential oversight for hydraulic fracturing
CERCLA–1980	None currently*/Clean-up of hazardous waste sites	Might hold companies responsible for clean-up and damages due to releases of hazardous materials at well sites and require reporting of toxic chemicals used in the fracturing process.
CWA–1972	EPA/Waters of the United States	NPDES stormwater permit required for discharges from well sites but could be extended to apply to temporary holding pits.
RCRA–1976	None currently*/Hazardous wastes	Could require the listing of hazardous substances used in the injection fluids in addition to regulation of the resulting wastewater flowback.
SDWA–2005 amendment	None currently*/Drinking water of the United States	The UIC program could regulate subsurface emplacement fluids that would include injection for gas development and underground storage of waste fluids.
SRBC–1971 and DRBC–1961	Commissioners/Susquehanna and Delaware River Basins	Regulates deposits or withdrawals from the river basin so that fracturing operations need permits to withdraw water for injecting into wells or for depositing wastewaters back into the river system.

*Note: * Exemptions exist that prohibit EPA from applying these standards to oil and gas extraction.*

Congress acted to protect drinking water in the Safe Drinking Water Act of 1976 with protection through the implementation of an Underground Injection Control program regulating subsurface injections and storage of fluids. But, in the Energy Policy Act of 2005, Congress enacted an exclusion to this program.

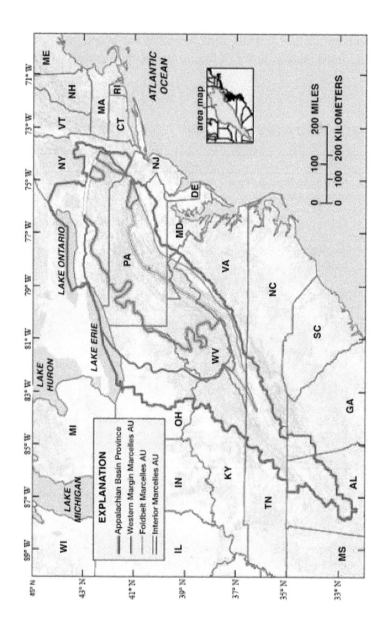

FIGURE 1: Map of the Marcellus shale assessment units (AU) which are located within the Appalachian Basin Province.

The term "underground injection"—(A) means the subsurface emplacement of fluids by well injection; and (B) excludes—(i) the underground injection of natural gas for purposes of storage; and (ii) the underground injection of fluids or propping agents (other than diesel fuels) pursuant to hydraulic fracturing operations related to oil, gas, or geothermal production activities [14].

While the Safe Drinking Water Act specifically excludes hydraulic fracturing from regulation, the use of diesel fuel in fracturing is regulated since it is defined as a hazardous contaminant [14].

Congress regulated hazardous waste from inception to disposal under the Resource Conservation and Recovery Act (RCRA) and EPA has developed a list of regulated substances [15]. However, RCRA does not regulate hazardous wastes involved in oil and gas extraction and production under RCRA Subtitle C. These materials are subject to state regulation under the less stringent RCRA Subtitle D solid waste regulations as well as other federal regulations, although states are also free to adopt more demanding provisions. In a publication regarding the exemption EPA says, "Although they are relieved from regulation as hazardous wastes, the exemption does not mean these wastes could not present a hazard to human health and the environment if improperly managed" [16].The absence of any federal requirement to disclosure hazardous chemicals used in fracturing is a major issue [17].

Hydraulic fracturing, like any deep drilling operation, is subject to the risk of leaks and spills that can cause areas to be contaminated by hazardous waste. In 1980, the Comprehensive Environmental Response, Compensation, and Liability Act (CERCLA) provided for the clean-up of abandoned hazardous waste and established liability to those who released the wastes to pay for clean-up [18]. Yet oil and gas exploration is exempt from clean-up of accidental spills, leaks, and problems from underground injection via the Energy Policy Act of 2005 [19]. Exploration and production companies cannot be held liable for damages under CERCLA, nor may they be sued by any entity for replacement of drinking water supplies or any health problems created as a result of their operations [20].

Applicable to fracturing regulation are two regional commissions that have jurisdiction over all water withdrawals from specific watersheds:

the Delaware River Basin Commission and the Susquehanna River Basin Commission. Figure 1 illustrates the overlap of the Marcellus shale formation and several river basins. Because of regulations adopted by these commissions, all oil and gas production operations must obtain permits before they can pump millions of gallons of water to use in their wells. Therefore, these commissions play a critical role in the continuation of oil and gas development in the Marcellus shale region because hydraulic fracturing cannot occur without significant quantities of water.

The Delaware River Basin Commission is a regulatory body that was established in 1961 by a congressional compact. It includes a division engineer from the US Army Corps of Engineers and representatives from New York, Pennsylvania, New Jersey, and Delaware who are appointed individually by the executive office in each state [21]. Any decision of the Commission involves the approval of all members. The Commission has full water resource management authority, including water allocations and diversions. Any project that will withdraw or discharge water in or from the basin must be approved by a process that includes a public hearing. In 2009, the Delaware River Basin Commission banned new exploration and production of shale gas in the region until strong regulations are in place. Public comments on draft regulations for natural gas well pad projects were closed in April 2011, and they are currently being reviewed by the commission [22].

Similarly, the Susquehanna River Basin Compact established the Susquehanna River Basin Commission, another federal-interstate regulatory collaboration by Congress and the member states. It is parallel in structure and authority to the Delaware River Basin Commission. Any decision of the Commission involves the approval of all of the member parties, which include the states of Maryland, Pennsylvania, and New York, as well as the federal government. The Susquehanna River is in the Marcellus shale region so any hydraulic fracturing operation using surface waters will need a permit (see Figure 1). At the 15 March 2012 commission meeting, several natural gas drilling projects were rejected and many more reconsidered or tabled [23].

Due to exceptions to federal environmental laws detailed above, the federal government does not have a clear role to play in the regulation of hydraulic fracturing as a result of amendments made to the environmental laws detailed above. The exceptions have allowed for more liberal oil and gas development in areas not within the Delaware and Susquehanna river basins. However, EPA has announced that new federal standards for fracturing wastewater are being developed [24]. Even if these are implemented, the regulatory authority to address potential risks has been passed down to the states. States in the region have a range of different approaches to address environmental concerns that accompany horizontal drilling.

9.3 STATE POLICIES AND ACTIONS

Five states in the Marcellus Shale region, New York, Pennsylvania, Ohio, West Virginia, and Virginia (Figure 1), have different approaches to regulating oil and gas development involving horizontal drilling. These distinct regulatory structures, as well as the significant policy changes made by states in the last two years, illustrate a spectrum of possibilities and outcomes. These structures and outcomes can be used to guide future policy alternatives and decisions. A summary of the current policies and incidences can be found in Table 2.

New York placed a moratorium on hydraulic fracturing in 2008 [25], and subsequently an executive order directed the state Department of Environmental Conservation to conduct a review and analysis of horizontal hydraulic fracturing [26]. The state has allowed hydraulic fracturing [27]; its horizontal hydraulic fracturing that is precluded. The state has developed a complex and comprehensive regulatory framework [28]. New York is also unique due to the local court battles between citizens and oil and gas companies concerning bans on hydraulic fracturing. Since 2008, 22 cities have rezoned to prohibit fracturing [29]. The city of Dryden is one of these local governments that banned horizontal hydraulic fracturing via a zoning law [30]. A natural gas production company filed suit against the city claiming that the municipality was overstepping its jurisdiction. In early 2012, a state superior judge ruled that the municipality was not preempted by state laws and had the right to tighten its land use regulations. Zoning bans by local governments across New York send a strong message about local disapproval of hydraulic fracturing and have established a precedent for other municipalities to limit the drilling rights of oil and gas companies. In New York, either a State Pollution Discharge Elimination System (SPDES) individual or general permit is required for fracturing activities that cause a discharge into surface waters [31]. For high-volume hydraulic fracturing activities, a special general permit has been proposed [32]. The special general permit addresses drilling operations from the construction phase through to the production phase, including well-site construction, soil disturbance, and potential contamination [32]. Hydraulic fracturing would be precluded in the New York City and Syracuse watersheds, on certain state lands, within 610 m of public drinking water supplies, and within 152 m of private wells. Furthermore, provisions require the identification and evaluation of fracturing fluid additives "to encourage the use of processes and substances that minimize the potential for environmental impacts" [32].

TABLE 2: Summary of state regulations and outcomes.

State	Regulatory Authority	Legislative Actions	Reported Incidents
New York	Department of Environmental Protection	• statewide moratorium • SPDES permit • disclosure of fracturing fluids • municipal zoning bans	• well water contamination from vertical wells
Ohio	Department of Natural Resources	• restrictions for impoundment pits • fees for wastewater disposal • electronic tracking	• earthquakes • well and surface water contamination
Pennsylvania	Department of Environmental Protection	• Act 13 of 2012 set stricter standards for oil and gas production and preempted most municipal regulations	• local water supply contamination
Virginia	Department of Mines Minerals and Energy	• the Gas and Oil Act allows the non-disclosure of chemicals and preempts municipal regulations	• noxious fumes • light pollution • well and surface water contamination • disruption from truck traffic
West Virginia	Department of Environmental Protection	• the Horizontal Well Act of 2011 preempted municipal regulations and exempts activities	• drinking water contamination and neurological disease • creek contamination with a massive fish kill

In Ohio, public concern about hydraulic fracturing came to a head on New Year's Eve 2011 when there was a 4.0 earthquake near the city of Youngstown. This seismic event followed several other earthquakes that began in March 2011, just three months after a 2804 m wastewater well was drilled in Youngstown for the storage of fracturing fluids. There is no record of seismic activity in this area during the previous 235 years [33]. Rather, the exponential growth in natural gas and storage well drilling in the area, jumping from an average of four new permanent-waste storage wells per year from 1990–2010 to 29 new wells in 2011, seems to have caused the seismic activity. Experts at the Ohio Department of Natural Resources concluded that the seismic disruptions were a result of brine injection related to hydraulic fracturing. Other research supports these claims, as it is widely understood that injecting fluid underground at high pressure can trigger earthquakes [34]. Any geologic disruption in an area where hazardous waste is permanently stored might result in contamination of ground and surface waters. In the last two years, Ohio has implemented further restrictions on impoundment pits located in urban areas, fees for disposal of wastewater via injections in wells, requirements for more comprehensive geologic data prior to permitting, and electronic tracking systems to identify the makeup of drilling wastewater fluids [35,36]. The Ohio Department of Natural Resources has regulatory authority over hydraulic fracturing activities in the state.

In Pennsylvania, the Department of Environmental Protection has authority over hydraulic fracturing activities. In 2012, the state legislature passed Act 13 containing stronger and more detailed regulations including increased setback requirements for unconventional gas development, enhanced protection of water supplies, and strong, uniform, consistent statewide environmental standards [37]. This legislation included a uniformity provision that attempted to preempt "all local ordinances regulating oil and gas operations" [38]; however, a Pennsylvania court found this provision to be unconstitutional [39].

The legislature of Virginia decided to encourage the economical extraction of Virginia's coalbed methane [40]. Through the Gas and Oil Act, the state legislature preempted local regulations to give the Virginia Department of Mines, Minerals, and Energy the exclusive authority to regulate activities relating to oil and gas exploration and production [41]. The act establishes regulations and permitting requirements that govern mineral extraction. The act does not require the reporting of the chemical composition of fracturing fluids.

In West Virginia, a number of environmental problems have allegedly been caused by hydraulic fracturing activities, as documented by Earthjustice and mapped on their website of "fraccidents" [42]. The Department of Environmental Protection (DEP) issued a proposal in 2010 to rewrite regulations for drilling operations across the state. After months of talks with various stakeholder groups, the state legislature adopted the Horizontal Well Act in 2011 [43]. While the act delineates requirements that should help protect the environment, a number of provisions limit this protection [44]. For example, the act sets forth exceptions so that vertical and permitted wells escape further regulation [44]. Wells disturbing less than three acres or using less than 200,000 gallons of water in a 30-day period are not subject to the requirements of the act [43]. Turning to local regulations, the act specifically provides that the secretary of the West Virginia Department of Environmental Protection "has sole and exclusive authority to regulate" activities related to hydraulic fracturing so that municipal governments cannot interfere with drilling [43]. Furthermore, the secretary has "broad authority to waive certain minimum requirements" if deemed appropriate [43].

9.4 POLICY ALTERNATIVES

The absence of comprehensive controls and differences of regulatory approaches to horizontal drilling employing hydraulic fracturing between states do not provide adequate protection of local and regional water resources. The legal battles and state legislative revisions in the Marcellus shale region indicate significant public concern about the safety of horizontal well drilling. With the introduction of many fractures along a horizontal wellbore, there are new risks to be considered [45]. Furthermore, these fracturing activities pose risks to river systems and water quality that do not recognize manmade state and municipal boundaries. An individual state is unable to preclude pollutants from upstream states so that multistate or federal controls become important for the maximization of social, environmental, economic, and democratic outcomes for the Marcellus shale region [46]. In a similar manner, a local government may not be in a position to maximize outcomes for a region. Rather, by directing its focus on a small geographic area, a municipality may overlook broader, regional concerns.

The analysis of federal, regional and state regulatory controls over horizontal drilling identify two options for reducing risks accompanying hydraulic fracturing. The first option involves deleting the oil and gas production exemption set forth by the Energy Act of 2005 and requiring disclosure of hazardous chemicals employed in hydraulic fracturing. By deleting the exemption for oil and gas exploration and production, provisions of the Safe Drinking Water Act would offer additional oversight to fracturing activities involving chemicals being injected into the ground. In addition, requiring mandatory reporting of chemicals used in hydraulic fracturing would allow first responders to blowout accidents and other mishaps to have sufficient information for selecting appropriate responses. States often lack adequate controls [47] and because the Marcellus shale formation spans multiple rivers and covers multiple states, a collective, standardized legal framework is needed to ensure equitable protection of the environment and to protect the economic interests of all parties involved.

In 2011, the Fracturing Responsibility and Awareness of Chemicals Act, which would repeal exemptions for hydraulic fracturing, was introduced in both houses of Congress [48]. However, the act remains in committees and, given the concern over rising energy prices, an initiative to restrict domestic energy production is an unpopular position for policy makers [49]. Moreover, the proposed FRAC Act does not require the public chemical disclosure requirements for fracturing fluids. Thus, the act fails to help identify sources of contamination that may occur from accidental releases and spills. States can enact requirements on the disclosure of chemicals, but most have chosen to include a provision for trade secret protection [50].

A second option to mitigate risks of water contamination by hydraulic fracturing is to strengthen safety controls for the disposal of flowback fluids. For drilling, damages from blowouts are a concern that can be addressed through better well construction standards and adequate construction monitoring and inspection [51]. For probabilistic events including unplanned accidents, the use of environmental impact assessments may reduce negative impacts [15] as well as inspections [52]. Moreover, since hydraulic fracturing in the Marcellus shale region leads to increased concentrations of Ra^{226}, Ra^{228}, and Ba in flowback waters from Marcellus wells [53], more definitive and demanding treatment specifications for fracturing fluids discharged to publically owned treatment works may be needed to allay concerns that downstream water users are being harmed.

9.5 CONCLUSIONS

In the absence of consistent federal standards, individual states, driven by their short-term interests, are allowing actions that lead to long-term damage to common resources. This allows firms to avoid costs reflected in the negative externalities of production. With respect to horizontal drilling, a state's interest is economic gain through liberal gas production without full consideration of regional river basins and ground water supplies [54]. The examination of legal structures regulating hydraulic fracturing provides numerous examples of negative impacts on water quality as a result of poor management of drilling activities. To strengthen the protection of water sources in the Marcellus shale region, federal regulatory exemptions for oil and gas exploration should be deleted and additional resources should be allocated to the management of environmental risks accompanying hydraulic fracturing.

In a similar manner, local governmental actions addressing horizontal drilling may not be optimal. Tension exists between state and local governments over the regulation of hydraulic fracturing because local prohibitions on drilling can thwart state objectives. While the historic delegation of duties and responsibilities to municipal governments enable these governments to take actions on matters of local concern, state legislatures are having second thoughts about whether horizontal fracturing activities are local. Given changes in technology, communications, and transportation, issues relegated to local governments over past centuries may no longer be local. Interconnections of jobs, commerce, and social structures among local governments create externalities that cannot be meaningfully addressed by an individual municipality. Local governments may constitute archaic divisions that create impediments to the well-being of people and the economy of a state. Thus, in exercising their sovereignty, state legislatures are acting to preclude local decisions regarding fracturing that interfere with overriding state objectives.

REFERENCES

1. US Environmental Protection Agency. Plan to Study the Potential Impacts of Hydraulic Fracturing on Drinking Water Resources; EPA/600/R-11/122/November; US Environmental Protection Agency: Washington, DC, USA, 2011.
2. Weinhold, B. The future of fracking. Environ. Health Perspect. 2012, 120, A272–A279.
3. Finkel, M.L.; Law, A. The rush to drill for natural gas: A public health cautionary tale. Am. J. Public Health 2011, 101, 784–785.
4. Furlow, J.D.; Hays, J.R., Jr. Disclosure with protection of trade secrets comes to the hydraulic fracturing revolution. Tex. Oil Gas Energy Law 2011, 7, 289–355.
5. US Energy Information Administration. Short-Term Energy Outlook, Table 5b: U.S. Regional Natural Gas Prices; US Energy Information Administration: Washington, DC, USA, 2012. Available online: http://www.eia.gov/forecasts/steo/tables/?tableN umber=16#mstartcode=2007 (accessed on 25 September 2012).
6. Rahm, B.G.; Riha, S.J. Toward strategic management of shale gas development: Regional, collective impacts on water resources. Environ. Sci. Policy 2012, 17, 12–23.
7. Wiseman, H. Risk and Response in Fracturing Policy. University of Colorado Law Review, 2013. FSU College of Law, Public Law Research Paper No. 594, Available online: http://ssrn.com/abstract=2017104 (accessed on 22 November 2012). in press.
8. Wiseman, H. State Enforcement of Shale Gas Regulations, Including Hydraulic Fracturing. Available online: http://ssrn.com/abstract=1992064 (assessed on 15 November 2012). Energy Institute, University of Texas White Paper, 25 August 2011; FSU College of Law, Public Law Research Paper No. 581.
9. DiCosmo, B. Jackson Downplays Concerns Over Broad EPA Oversight of Fracking Wells; Clean Energy Report; US Environmental Protection Agency: Washington, DC, USA, 2012.
10. US EPA. Summary of the Clean Water Act; US Environmental Protection Agency: Washington, DC, USA, 1972. Available online: http://www.epa.gov/lawsregs/laws/ cwa.html (accessed 25 September 2012).
11. Navigation and Navigable Waters. United States Code, Sections 1311 and 1362, Title 33, Supplement 5, 2006.
12. Technology-Based Treatment Requirements in Permits. US Code of Federal Regulations, Section 125.3, Title 40, 2011.
13. Obold, J. Leading by example: The Fracturing Responsibility and Awareness of Chemicals Act of 2011 as a catalyst for international drilling reform. Colo. J. Int. Environ. Law Policy 2012, 23, 473–500.

14. Regulations for State Programs. United States Code, Section 300H, Title 42, Supplement 4, 2006.

15. US EPA. Summary of the Resource Conservation and Recovery Act; US Environmental Protection Agency: Washington, DC, USA, 2012. Available online: http://www.epa.gov/lawsregs/laws/rcra.html (accessed on 25 September 2012).

16. US EPA. Exemption of Oil and Gas Exploration and Production Wastes from Federal Hazardous Waste Regulations; EPA530-K-01-004; US Environmental Protection Agency: Washington, DC, USA, 2002. Available online: http://www.epa.gov/osw/nonhaz/industrial/special/oil/oil-gas.pdf (accessed on 25 September 2012).

17. Wiseman, H. Trade secrets, disclosure, and dissent in a fracturing energy revolution. Columbia Law Rev. Sidebar 2011, 111, 1–13.

18. Comprehensive Environmental Response, Compensation, and Liability. United States Code, Sections 9601–9675, Title 42, Supplement 4, 2006.

19. Energy Policy Act of 2005. Public Law 109–58, Section 106, 2005.

20. McKay, L.K.; Johnson, R.H.; Salita, L.A. Science and the reasonable development of Marcellus shale natural gas resources in Pennsylvania and New York. Energy Law J. 2011, 32, 125–143.

21. Delaware River Basin Commission Home Page. Available online: http://www.state.nj.us/drbc/ (accessed on 25 September 2012).

22. Delaware River Basin Commission. Draft Natural Gas Development Regulations; Delaware River Basin Commission: West Trenton, NJ, USA, 2011. Available online: http://www.nj.gov/drbc/programs/natural/draft-regulations.html (accessed on 25 September 2012).

23. Susquehanna River Basin Commission. SRBC met March 15: Reconsidered 22 and approved 20 additional projects; Denied 3 applications; Released proposed low flow policy for public comment. Available online: http://www.srbc.net/newsroom/News-Release.aspx?NewsReleaseID=81 (accessed on 25 September 2012).

24. US Environmental Protection Agency. EPA Announces Schedule to Develop Natural Gas Wastewater Standards/Announcement is part of administration's priority to ensure natural gas development continues safely and responsibly. Available online: http://yosemite.epa.gov/opa/admpress.nsf/d0cf6618525a9efb85257359003fb69d/91e7fadb4b114c4a8525792f00542001!OpenDocument (assessed on 2 November 2012).

25. Applebome, P. Drilling Critics Face a Divide over the Goal of Their Fight. New York Times 2012, A17.

26. Governor's Office, New York. Executive Order No. 41: Requiring further environmental review. Available online: http://www.governor.ny.gov/archive/paterson/press/121110PatersonExecutiveO-HydraulicFracturing.html (assessed on 2 November 2012).

27. Nolon, J.R.; Polidoro, V. Hydrofracting: Disturbances both geological and political: Who decides? Urban Lawyer 2011, 44, 507–532.

28. Mergen, A.C.; Aagaard, T.; Baillie, J.; Bender, P.; Beauduy, T.W.; Engelder, T.; Perry, S.; Ubinger, J.W., Jr.; Wiseman, H. "Shale" we drill? The legal and environmental impacts of extracting natural gas from Marcellus shale. Villanova Environ. Law J. 2011, 22, 189–224.

29. Calfee, C.; Weissman, E. Permission to transition: Zoning and the transition movement. Plan. Environ. Law 2012, 64, 3–10.

30. Anschutz Exploration Corporation v. Town of Dryden. 940 N.Y.S.2d 458. Supreme Court of New York, Tompkins County, 21 February 2012.

31. New York State Department of Environmental Conservation. SPDES General Permit for Stormwater Discharges from High Volume Hydraulic Fracturing Operations, Permit No. GP-0-XX-XXX, Fact Sheet; New York State Department of Environmental Conservation: Albany, NY, USA, 2011. Available online: http://www.dec.ny.gov/docs/water_pdf/hvhfgpfactsht.pdf (accessed on 27 September 2012).

32. New York State Department of Environmental Conservation. Proposed express terms 6 NYCRR parts 750.1 and 750.3. Available online: http://www.dec.ny.gov/regulations/77383.html (accessed on 27 September 2012).

33. Ohio Department of Natural Resources. Catalog and maps of Ohio earthquakes. Available online: http://www.dnr.state.oh.us/geosurvey/html/eqcatlog/tabid/8302/Default.aspx (accessed on 26 September 2012).

34. Raleigh, C.B.; Healy, J.H.; Bredehoeft, J.D. An experiment in earthquake control at Rangley, Colorado. Science 1976, 191, 1230–1242.

35. Division of Mineral Resources Management—Oil and Gas. Ohio Administrative Code, Chapter 1501:9, 2011.

36. Ohio Department of Natural Resources. Preliminary Report on the Northstar 1 Class II Injection Well and the Seismic Events in the Youghstown, Ohio Area; Ohio Department of Natural Resources: Columbus, OH, USA, March 2012. Available online: http://ohiodnr.com/downloads/northstar/UICreport.pdf (accessed 25 September 2012).

37. Pennsylvania Department of Environmental Protection. Act 13 of 2012. Available online: www.portal.state.pa.us/portal/server.pt/community/act_13/20789 (accessed on 25 September 2012).

38. Uniformity of Local Ordinances. Pennsylvania Consolidated Statutes, Section 3304, Title 58, 2012.

39. Robinson Township v. Commonwealth of Pennsylvania, 52 A.3d 463. Commonwealth Court of Pennsylvania, 26 July 2012.

40. Miller, T.; Kauffman, M.K. Is It a Deep Well or a Shallow Well and Who Cares?; Energy & Mineral Law Institute: Lexington, Kentucky, USA, 2010; Volume 31, pp. 404–431. Chapter 12.

41. Exclusivity of Regulation and Enforcement. Code of Virginia, Section 45.1-361.5, Title 45.1, 2012.

42. Earthjustice. West Virginia and fracking. Available online: http://earthjustice.org/features/campaigns/west-virginia-and-fracking (accessed on 25 September 2012).

43. Legislative Findings; Declaration of Public Policy. West Virginia Code, Section 22-61-2, Chapter 22, 2012.

44. Garvin, D.S.; Coordinator, L.., Jr. WV Environmental Coalition. Legislature Passes Weak Marcellus Shale Bill in Special Session; Sierra Club: San Francisco, CA, USA, 2012. Available online: http://westvirginia.sierraclub.org/newsletter/archives/2012/03/a_001.html (accessed on 26 September 2012).

45. He, H.; Dou, L.; Fan, J.; Du, T.; Sun, X. Deep-hole directional fracturing of thick hard roof for rockburst prevention. Tunn. Undergr. Space Technol. 2012, 32, 34–43.

46. Vogel, E. Parceling out the watershed: The recurring consequences of organizing Columbia River Management within a basin-based territory. Water Alternat. 2012, 5, 161–190.

47. Wiseman, H.; Gradijan, F. Regulation of Shale Gas Development, Including Hydraulic Fracturing. Energy Institute, The University of Texas at Austin; 31 October 2011. University of Tulsa Legal Studies Research Paper No. 2011-11. Available online: http://papers.ssrn.com/sol3/papers.cfm?abstract_id=1953547 (assessed 15 November 2012).

48. US House of Representations Bill No. 1084, Fracturing Responsibility and Awareness of Chemicals Act of 2011. Available online: http://www.govtrack.us/congress/bills/112/hr1084 (accessed 27 September 2012).

49. Bittle, S.; Rochkind, J.; Bosk, J. Confidence in U.S. Foreign Policy Index: Energy, Economy New Focal Points for Anxiety over U.S. Foreign Policy; Public Agenda: New York, NY, USA, 2008. Available online: http://www.publicagenda.org/files/pdf/foreign_policy_index_spring08.pdf (accessed on 25 September 2012).

50. Murrill, B.J.; Vann, A. Hydraulic Fracturing: Chemical Disclosure Requirements; CRS Report for Congress; Congressional Research Service: Washington, DC, USA, 2012.

51. Zhang, J. Pore pressure prediction from well logs: Methods, modifications, and new approaches. Earth. Sci. Rev. 2011, 108, 50–63.

52. Wiseman, H. Fracturing legislation applied. Duke Environ. Law Policy Forum 2012, 22, 361–384.

53. Haluszczak, L.O.; Rose, A.W.; Kump, L.R. Geochemical evaluation of flowback brine from Marcellus gas wells in Pennsylvania, USA. Appl. Geochem. 2012. in press..

54. Powers, E.C. Fracking and federalism: Support for an adaptive approach that avoids the tragedy of the regulatory commons. J. Law Policy 2011, 19, 913–971.

CHAPTER 10

Excerpt from: Reflecting Risk: Chemical Disclosure and Hydraulic Fracturing

SARA GOSMAN

10.1 THE FAILURES OF FEDERAL CHEMICAL LAW

To understand why states, the oil and gas industry, and stakeholders turned to the policy approach of chemical disclosure, it is first necessary to understand why existing federal chemical law fails to address long-term uncertainty about the effects of hydraulic fracturing fluids. There are three federal laws that would seem to apply to the chemical activity: the Emergency Planning and Community Right-to-Know Act (EPCRA), [181] the primary federal law designed to inform the public about chemical risk; the Toxic Substances Control Act (TSCA), [182] the primary federal law governing the production and use of toxic chemicals; and the Safe Drinking Water Act (SDWA), [183] the primary federal law governing underground injection of fluids. [184] Yet the nature of the chemical activity poses a fundamental challenge to these laws, including the assumptions about risk

Reprinted with permission from the author and the Georgia Law Review. *The selection reproduced here is an excerpt from a larger report: Reflecting Risk: Chemical Disclosure and Hydraulic Fracturing. Gosman S.* Georgia Law Review *83 (2013). http://georgialawreview.org/wp-content/up-loads/2014/07/Download-PDF-V48-I1-Gosman.pdf.*

and the need for trade secret protection on which these laws are based. While high-volume hydraulic fracturing is not unique in presenting this challenge, the scope of the activity places the laws' failures in high relief.

10.1.1 THE EMERGENCY PLANNING AND COMMUNITY RIGHT-TO-KNOW ACT

The chemical activity of high-volume hydraulic fracturing fits poorly within the structure of the EPCRA. Because hydraulic fracturing fluids are mixtures of many constituents in very small concentrations, released into the environment by multiple sources, the fluids are rarely subject to the law's reporting requirements. The two "right-to-know" aspects of the law—the TRI and the emergency planning provisions—focus on the risks posed by large amounts of a single chemical at one site. Small concentrations of constituents in a chemical product are ignored, regardless of the total amount and toxicity. [185] And because manufacturers of chemical additives must invest resources in research and development to compete in the evolving market, they are likely to claim trade secret protection for active chemicals in the event that reporting is required.

Oil and gas well operators are not required to report any releases of toxic chemicals to the TRI. Neither Congress nor the EPA has chosen to include the industry in the list of industries that must submit reports. [186] This decision cannot be attributed solely to a failure of political will. A well site is unlikely to meet the threshold of ten or more full-time workers and use of more than 10,000 pounds of a reportable chemical in a year. [187] Even if a service company hydraulically fractures multiple horizontal wells on a site, the site is still unlikely to meet the threshold for any one chemical. [188] This is true although the current list of 682 reportable chemicals and chemical categories includes several chemical constituents found in hydraulic fracturing fluids, [189] each of which is "released" into the environment by injection. [190] While the EPA could use its administrative authority to aggregate the chemical activities of multiple well sites in an area, it has so far demurred. [191]

Well operators are also generally not required to disclose hazardous chemicals on a well site to state and local emergency authorities and the public. Under the EPCRA's emergency planning provisions, an operator must submit the name of a hazardous chemical [192] or a Material Safety Data Sheet (MSDS) and an annual inventory if there are 10,000 pounds of the chemical at the site at one time. [193] Extremely hazardous chemicals are subject to a lower threshold. [194] Although many hazardous [195] and extremely hazardous chemicals [196] are found in high-volume hydraulic fracturing fluids, it is unlikely that a single chemical additive product or constituent would be present on the site in the threshold quantity. [197] This is in part because the well operator may choose to treat each hazardous constituent of a product as a separate chemical for purposes of reporting. [198] There is one exception to the threshold requirement. A member of the public may obtain an inventory of hazardous chemicals present in less than the threshold amounts upon a statement of "general need." [199]

If a well operator were required to submit information on hazardous chemicals, this information would be limited by manufacturers' prior trade secret claims under worker safety law. The Hazard Communication Standard allows a manufacturer of a mixture to protect the name of a chemical, other specific identification, and the exact concentration of a constituent from disclosure. [200] The manufacturer need not justify the claim to the Occupational Safety and Health Administration, and there is no process for the general public to challenge the claim. [201] The EPA does not review the claim when the information is required under the EPCRA. Only health professionals and employees may obtain the identity of the chemical; with the exception of medical emergencies, these professionals and employees must show an occupational health purpose, explain why other information is not sufficient, and sign a confidentiality agreement. [202]

In the unlikely event that the operator of a well site wanted to withhold additional chemical information, the EPCRA provides much more limited protection for trade secrets. For example, the operator may only withhold the specific identity of the chemical on site. [203] Thus, the chemical's exact concentration in a mixture, the amount at the site, and the generic class or category of the chemical must still be reported. [204] The operator must also justify the trade secret claim, and share an "unsanitized" version of the chemical list and inventory. [205] Any person may petition the EPA to review the trade secret claim; if the EPA finds that the claim is unjustified, the agency will disclose the information. [206] In addition, the state is required to provide information about the adverse health effects of proprietary chemicals to "any person requesting information" about the substance. [207] None of these more stringent requirements apply to the trade secret claims of manufacturers of chemical additives, however. [208]

10.1.2 THE TOXIC SUBSTANCES CONTROL ACT

The chemical activity of high-volume hydraulic fracturing also fits poorly within the structure of the TSCA. Recognizing that information is necessary for effective chemicals regulation, Congress gave the EPA broad rulemaking authority to gather data on chemicals and mixtures, [209] and also imposed duties on companies to record and report what they know. [210] With a few exceptions, however, the Agency has used this authority to target large manufacturers of individual chemicals rather than the manufacturers or processors of mixtures. [211] The TSCA encourages the EPA to focus on the risks of chemical constituents apart from the risks of hydraulic fracturing fluid, and on manufacturing and processing the chemical additives rather than on the end use of high-volume hydraulic fracturing. Paradoxically, the absence of information on hydraulic fracturing fluids and the preference for regulating chemicals over mixtures makes it difficult for the EPA to require companies to test the fluids for health and environmental effects. [212]

The EPA faces several hurdles in gathering information about hydraulic fracturing fluids and their effects. At the outset, it is unclear whether the EPA can require service companies to provide data on hydraulic fracturing fluid itself. [213] The Agency can obtain indirect information on the fluid from manufacturers of chemical additive mixtures, but only to the extent that the information "is necessary for the effective enforcement of [the TSCA]." [214] The EPA is also prohibited from obtaining information "with respect to changes in the proportions of the components" of chemical additives unless it makes the same determination. [215] Neither the courts nor the EPA have determined which information is necessary for effective enforcement. In response to a petition by environmental organizations, the EPA agreed in 2011 to initiate a rulemaking to obtain existing data on chemicals and mixtures used in hydraulic fracturing. [216] This data could include, for example, the identity, categories of use, and effects of chemical additives, and submission of existing health and safety studies. [217] As of late 2013, the EPA has not issued an advanced notice of proposed rulemaking.

While manufacturers and processors of chemicals or mixtures must maintain records on significant adverse reactions and immediately report substantial risks of injury to health or the environment, [218] these requirements are unlikely to provide significant amounts of information on hydraulic fracturing fluids. As above, it is unclear whether the TSCA requires service companies to gather information on fluids. [219] Companies that manufacture and process chemical additives are subject to the requirements, but only severe effects that are unknown to the federal government must be recorded or reported. [220] In addition, there is no requirement to investigate incidents or go beyond what the company reasonably knows. [221] The EPA has neither required submission of significant adverse reaction records on chemical additives, nor disclosed any information it has received on substantial risks of the additives.

If companies were required to submit information on additives or fluids to the EPA, the TSCA provides relatively broad protection for trade secrets and confidential commercial information. [222] In contrast to the requirement under worker safety law, however, a company must submit the confidential information to the EPA and provide a written justification. [223] Protected information may include the identity of a chemical, the composition of a mixture, the location of the chemical activity, and other information on processing and use. [224] The primary exception is health and safety studies, which the TSCA specifically exempts from confidentiality claims if no information on processes or the proportions of mixtures is released. [225] Until recently, companies could claim that a chemical identity was a trade secret because other companies could obtain process information from the chemical name. [226] In 2010, however, the EPA announced that it would begin to scrutinize these claims more carefully. [227] Unlike the EPCRA, [228] there is no process by which the public may challenge the validity of the company's claims.

10.1.3 THE SAFE DRINKING WATER ACT

Finally, the chemical activity of high-volume hydraulic fracturing fits poorly within the structure of the SDWA. Hydraulic fracturing is specifically exempt from the statute's Underground Injection Control (UIC) program unless the fluid contains diesel fuels. [229] Therefore, most oil and gas well operators need not obtain a UIC permit and meet minimum construction, operating, and monitoring requirements so that the fluids will not endanger drinking water sources. [230] But even if all operators were required to obtain a permit, the UIC program would not necessarily provide information on the risks of a mixture comprised of many constituents in very small concentrations.

The EPA's existing testing and monitoring requirements gather very little information on the potential health and environmental effects of complex mixtures. The UIC program divides wells into six classes based on the expected risk of the injected fluids. [231] Operators of "Class II" wells, which inject fluids associated with oil and gas production, [232] are only required to provide information on certain physical and chemical characteristics of the fluids, such as pH, major ions, and trace elements. [233] Even operators of "Class I" wells, which accept industrial and municipal wastewater, are not required to provide comprehensive toxicity data. [234] The toxicity test to determine if the proposed injectate is hazardous detects forty-three substances. [235] In the EPA's Region 5, the operator must test for all hazardous constituents that comprise a "major portion" of the waste stream, or greater than 0.01% by mass. [236] Most constituents of hydraulic fracturing fluid would not meet this threshold. [237]

Perhaps cognizant of the limitations in its existing program, the EPA has proposed using the policy approach of chemical disclosure to gather information on hydraulic fracturing fluid containing diesel fuels. [238] In a 2012 draft guidance, the EPA recommends that a well operator submit a "detailed chemical plan describing the proposed fracturing fluid composition, including the volume and range of concentrations for each constituent," with the permit application. [239] The EPA also recommends that monitoring requirements include "fracturing fluid composition data." [240] Operators may claim that the data are a trade secret; the statute generally protects trade secrets and "secret processes" from public disclosure after a showing "satisfactory" to the EPA. [241] The EPA extended the comment period on the guidance because of significant interest in—and criticism of—the recommendations. [242] The Agency has not issued any further updates since then.

10.2 THE ADVENT OF STATE CHEMICAL DISCLOSURE

The same assumptions about risk and the need for trade secret protections, combined with a faith in engineering controls, have historically led states not to regulate the chemicals used in highvolume hydraulic fracturing. After the U.S. Congress considered a bill to repeal the SDWA exemption and require disclosure of chemical constituents, however, the states rushed to adopt chemical disclosure policies. All of the states focus on the very aspects of the activity that pose a challenge to federal laws: the risks of a mixture comprised of many constituents in very small concentrations.

10.2.1 THE RUSH TO DISCLOSE

States have traditionally regulated oil and gas wells through permitting programs that govern the drilling, completion, production, and plugging and abandonment of oil and gas wells. [243] The laws seek to protect water resources by specifying the process of drilling, the integrity of well construction, and the proper plugging and abandonment of wells. [244] For example, state regulations generally prescribe the material, placement, and number of casing strings, and how the casing is to be cemented. [245] In a 2009 survey, the Groundwater Protection Council (GWPC), a non-profit organization of state UIC and groundwater program officials, [246] found that almost all of the twenty-seven then-producing states required "surface casing" to run from the ground surface to below the deepest groundwater zone, and to be fully cemented to ensure that fluids cannot escape. [247] In addition, most states prescribed either a "waiting period" for the cement to cure or a cement integrity test. [248]

Until 2010, the oil and gas producing states primarily regulated the technique of hydraulic fracturing by requiring operators to submit a general report between thirty and sixty days after well completion. [249] Of the twenty-seven states surveyed by the GWPC, twenty-five required such a report. [250] The reports most often asked for information about the treatment depth, materials, and volumes of fluid. [251] Ten of the states required some general information about the hydraulic fracturing fluid, but none asked for a list of chemical constituents. [252] A few states went beyond reporting to require operators to obtain approval of the hydraulic fracturing treatment. [253] Wyoming specified that the fluids be designed "to prevent significant dissolution" of trona, a sodium-rich mineral found in parts of the state. [254] Yet even in these states, there was no review and approval of the specific chemical constituents to be used. Other states simply prohibited hydraulic fracturing from causing pollution, [255] or sought to control damage after the fact by requiring operators to notify the Agency and to proceed diligently to repair the damage. [256]

In June 2009, U.S. Representative Diana DeGette (D-CO) and U.S. Senator Bob Casey (D-PA) introduced the Fracturing Responsibility and Awareness of Chemicals (FRAC) Act in Congress. [257] The FRAC Act would have repealed the exemption for hydraulic fracturing in the SDWA and preempted state laws governing oil and gas wells. [258] The Act would also have required "any person using hydraulic fracturing" to report the chemical constituents in the fluid to the state or the EPA. [259] In turn, the state or the EPA would post the constituents on "an appropriate Internet website." [260] In the case of a medical emergency, the Act would have required proprietary information on formulas or the specific identity to be released. [261]

The states took swift action to counter this threat to their authority. Some states chose to regulate several different aspects of hydraulic fracturing; others just a few. [262] Some chose to include all types of hydraulic fracturing; others focused on high-volume hydraulic fracturing. [263] But all of the state policies required some form of chemical disclosure. [264] In August 2010, Wyoming became the first state to adopt a disclosure policy. [265] Arkansas followed suit that December. [266] In 2011, the pace accelerated. Ten more states—Colorado, [267] Idaho, [268] Indiana, [269] Louisiana, [270] Michigan, [271] Montana, [272] New Mexico, [273] Pennsylvania, [274] Texas, [275] and West Virginia [276]—adopted policies. New York, which has placed a moratorium on hydraulic fracturing until it issues regulations, also released the first draft of its disclosure requirement. [277] In 2012, the pace accelerated even more. Four states— North Dakota, [278] Ohio, [279] Oklahoma, [280] and Utah [281]—adopted policies, the Pennsylvania General Assembly enacted an additional policy, [282] and Indiana extended disclosure to all wells. [283] The Louisiana Legislature also enacted a policy that largely mirrors the 2011 rule. [284] Regulatory agencies in an additional six states— Alaska, [285] California, [286] Mississippi, [287] Nebraska, [288] South Dakota, [289] and Tennessee [290]—proposed policies. New York released a second draft of its proposed regulations. [291] And legislatures in two more states—Kansas [292] and North Carolina [293]—gave their regulatory agencies specific authority to adopt policies. In an ironic twist, the U.S. Bureau of Land Management (BLM), which is responsible for federal mineral interests as well as Indian mineral interests, proposed a disclosure policy similar to the state policies in May of that year. [294]

The speed at which chemical disclosure has spread across the nation is stunning. There are thirty-five states in which hydraulic fracturing is or could be occurring. [295] In just three years, thirty of these states have considered a disclosure policy, and twenty-two states have policies in place. As of late 2013, California, [296] Illinois, [297] Mississippi, [298] Tennessee, [299] Nevada, [300] and South Dakota [301] have final policies. West Virginia has replaced its emergency policy with a permanent one. [302] Alaska [303] and Nebraska [304] have revised their proposed regulations after public comment; Kansas has proposed a policy; [305] Michigan has proposed a new policy; [306] and North Carolina is in the process of proposing a policy. [307] Maryland has released a draft report that recommends a policy. [308] In the most recent session, the Florida Legislature considered two bills on disclosure, one of which passed the Florida House but died in the Senate. [309] Arizona has encouraged operators to voluntarily disclose constituents. [310] Finally, the BLM has revised its proposed regulation. [311] Only Alabama, [312] Kentucky, Missouri, Oregon, and Virginia have not taken action. No other environmental issue in recent memory has spurred such a fast and uniform response by states.

There are many possible reasons for this turn of events. No doubt, the fear that the federal government would usurp the states' traditional authority over oil and gas wells pushed the states to act quickly. [313] Public pressure to reveal the chemical constituents was strong, and the oil and gas industry supported disclosure in the hope that it would calm the grow-ing controversy. [314] But the policies were also relatively easy to adopt. Almost all producing states already required well operators to report infor-mation on the treatment. Moreover, disclosure fit comfortably within the educational efforts taken by states to address public concern. For the many state officials that were surprised by the public's reaction to high-volume hydraulic fracturing, the policies provided an opportunity to continue edu-cating citizens about the "real" risks and to explain why the existing per-mitting programs protected water resources. [315]

10.2.2 THE THREE MODELS OF DISCLOSURE

All of the state policies are comprised of five elements: (1) the object of disclosure shares (2) the subject chemical information on the total hydrau-lic fracturing fluid (3) through a specific means (4) to the designated audi-ence (5) while ensuring the confidentiality of trade secrets. [316] Using these elements, the policies can be divided into three models: the reporting model, the regulatory model, and the public education model. Each model utilizes a distinct combination of elements. Some states have adopted one model; others have adopted two.

The reporting model follows the states' existing approach to chemical use. The primary purpose of the policy is to create arecord on each well, so that the state can respond to an incident involving the fluids. The state may also use the information to track trends in chemical use over time. In a typical policy, the well operator—the object of disclosure—must submit additional information on chemicals in the fluid to the state through the means of filing a well completion report after the treatment. The subject of disclosure ranges from MSDSs to lists of chemical constituents and their concentrations. Because the audience is the state, the policy does not necessarily address how to communicate information to the public. The state may post some information on its website, but the public generally accesses the reports through public records law or by searching an online version of the state's regulatory database. [317] The state is also likely to rely on established exceptions for trade secrets, whether in its governing oil and gas law or in public records law.

FIGURE 1: State Disclosure Models

Pennsylvania's initial disclosure policy uses the reporting model. Within thirty days after completion, the operator of a hydraulically fractured well must submit a stimulation record to the state as part of the well completion report. [318] The record includes a descriptive list of the chemical additives and the concentration of each additive in the fluid. [319] For every additive, the operator must provide a list of the hazardous constituents in the product MSDS, the chemical names, and the unique Chemical Abstracts Service (CAS) numbers. [320] The operator is also required to submit the concentration of each constituent. [321] Non-hazardous constituents need only be disclosed if the state makes a written request. [322] There is no provision for public disclosure; the public may view the information by making a records request or by inspecting the records in a state office. [323] Information in the report may be designated as a trade secret or as confidential, and the state then determines whether to release the designated information to the public using the state's Right-to-Know Law. [324]

The regulatory model expands the states' regulatory program. The primary purpose of the policy is to inform agency decisionmaking so that regulators can weigh the risks, decide whether to approve the activity, and ensure compliance. In a typical policy, the well operator—the object of disclosure—must submit additional information on chemicals in the fluid to the state through the means of permitting forms and reports. The subject of disclosure varies, but always includes information on the proposed chemical use and information on the actual chemicals used. Because the audience is the state, the public generally finds the information through records requests or online state databases. If confidentiality provisions are included in the policy, the operator may be allowed to assert broad trade secret claims as long as all information is given to regulators.

Wyoming, the first state to adopt a disclosure policy, uses the regulatory model. Before hydraulically fracturing a well, the owner, operator, or service company must provide to the state, "for each stage of the well stimulation program, the chemical additives, compounds and concentrations or rates proposed to be mixed and injected." [325] Each additive must be listed by function, such as acid or biocide, and by proposed rate or concentration. [326] The chemical constituents of the additive must also be identified by chemical name and the unique CAS number. [327]

Prior approval is required for the use of volatile organic compounds such as benzene, toluene, ethylbenzene, and xylene, and petroleum distillates. [328] Within thirty days after hydraulic fracturing, the owner or operator must report the details of each stage of the actual treatment, including the name, function, concentration or rate, and amount of each chemical additive and chemical constituent. [329] An owner or operator may request that the state provide confidentiality protection to proprietary information under the state's public records law. [330] There is no provision for public disclosure; the public may access the information through a records search of the database on the state website. [331]

Finally, the public education model discards the traditional tools of reporting and permitting and offers information directly to the public. The primary purpose of the policy is to educate individuals who are concerned about risk. In a typical policy, the service company or chemical supplier— the object of disclosure— must give additional information on chemicals in the fluid to the well operator, who in turn provides the information directly to the public. The state is now a facilitator and translator, not a regulator. The data are online, easily accessible, and presented together with explanations of the key issues. The subject of disclosure varies, but at a minimum includes the identity and maximum concentration of hazardous chemical constituents, as defined in worker safety law. When an operator withholds information because it is confidential, the policy requires the operator to clearly state that it is doing so. The policy may also limit trade secret protection by requiring disclosure of other identifying information or by ensuring that health professionals receive information when there is an incident involving hydraulic fracturing fluid.

Texas was the first state to adopt a policy using the public education model. After hydraulically fracturing a well, the service company or supplier must provide the well operator with information about the chemical use. [332] The operator is then required to submit the information within thirty days of completion to an independent website known as FracFocus. [333] FracFocus is managed by the GWPC and the Interstate Oil & Gas Compact Commission (IOGCC), composed of the governors of the oil and gas producing states. [334] For each chemical additive, the operator must disclose the trade name, the name of the supplier, and a brief description of the intended use or function. [335] The operator must also provide a list of the chemical constituents of the additives, including the CAS numbers, and the actual or maximum concentrations of hazardous substances. [336] No information is provided directly to the state unless FracFocus is inoperable. [337] On FracFocus, the public may browse the mapbased interface for a well; each well is linked to a standardized information sheet. [338] Texas's policy includes detailed trade secret provisions, including a process for nearby landowners to challenge claims, and exceptions for health professionals and emergency responders based on those in worker safety law. [339] If the identity of a constituent is withheld, the operator must disclose the chemical family name or a similar description on Frac-Focus. [340]

As can be seen in Figure 1, the vast majority of states that have adopted or proposed a disclosure policy have chosen the public education model pioneered by Texas. Of the states that use one model, nine states—Colorado, [341] Louisiana, [342] Nevada, [343] North Dakota, [344] Ohio, [345] Oklahoma, [346] South Dakota, [347] Texas, [348] and Utah [349]—have public education policies that either require or allow reporting to FracFocus. Four more states—Kansas, [350] Michigan, [351] Nebraska, [352] and North Carolina [353]—have proposed such policies. Of the states that use a combination of models, six states—California, [354] Mississippi, [355] Montana, [356] Pennsylvania, [357] Tennessee, [358] and West Virginia [359]—have adopted public education policies that require or allow reporting to FracFocus, and two states—Alaska [360] and New York [361]—are considering doing so.

Only Arkansas [362] and Illinois [363] have chosen to adopt public education policies that use state websites as the sole means of disclosure.

10.2.3 THE INFLUENCE OF FRACFOCUS

Because of the popularity of the public education model, FracFocus is not just a means of disclosure. The standardized reporting requirements that apply to all operators who submit data to the site, and the manner in which information is presented on the site, have, in effect, created a nationwide disclosure policy built on the public education model. Some states explicitly recognize this policy by deferring to FracFocus to specify the information required; [364] other states attempt to control the policy by threatening to use other means of disclosure if the site is not improved. [365]

The FracFocus policy can result in more information than would otherwise be provided. Many operators began to report information to FracFocus before the states adopted disclosure policies. In the first year of the site, operators reported half of all hydraulically fractured wells. [366] Operators in states that do not currently have a disclosure policy are also reporting. [367] In some states that do have a disclosure policy, operators must submit additional information such as the maximum concentration of each ingredient in the additive. [368] And in states that have only a regulatory or reporting policy and do not make information easily available, FracFocus provides a simple means for the public to access information that is also submitted to the state. [369]

The FracFocus policy can also result in less information, by in practice preempting more stringent state policies. [370] In Ohio, for example, an operator is required to report the total volume of the hydraulic fracturing fluid and the maximum concentration of each additive in the fluid to FracFocus. [371] Operators do not appear to be complying with the requirement because the standardized form does not call for this information. [372] Some states require an operator that withholds the identity of a constituent as a trade secret to meet the applicable standard in state law and to provide the chemical family or a similar descriptor. [373] [But FracFocus states that proprietary information is to be withheld using the standard in worker safety law, which does not require other identifying information to be disclosed. [374] In Texas, an operator must provide the name and contact information of the person making the trade secret claim on FracFocus, yet once again the FracFocus form does not include a place for the information. [375]

Moreover, the structure of FracFocus makes it difficult for states to ensure compliance with disclosure requirements. All laws are challenged by compliance, and there is no easy means of determining whether the extent of non-compliance on FracFocus is better or worse than in other contexts. But the sheer number of wells—as of late 2013, the site contains data on almost 56,000 wells [376]—makes careful oversight improbable. The entities that operate the site, the GWPC and the IOGCC, do not have authority to enforce state policies, and it is unclear to what extent states are aware of submissions. [377] Compliance is particularly an issue for trade secret claims, since they directly limit the amount of information. Of the states that use FracFocus, only California reviews and approves the claims. [378]

10.3 RE-ENVISIONING CHEMICAL DISCLOSURE

Taken together, the state policies provide much more information on the chemicals used in high-volume hydraulic fracturing than federal law and prior state laws. Indeed, this Article has relied on information generated by the policies to explain the chemical activity. But the amount of information does not necessarily measure the success of the policy in responding to long-term uncertainty about the health and environmental effects of the activity. By the standards of risk science and decision science, the FracFocus policy fails to meaningfully improve public understanding of risk. States should heed the lessons of the two fields and take a new approach to disclosure, one that combines additional data with neutral risk communications.

10.3.1 LEARNING FROM RISK SCIENCE

From the perspective of risk science, state disclosure policies meaningfully improve public understanding if risk assessors can ultimately characterize the cumulative risks of the chemical activity. By this logic, policies must at a minimum provide complete, detailed information on the composition of the fluids. The information must be presented in a way that allows experts to easily compare chemical compositions across wells. When default assumptions cannot resolve critical second-order uncertainties in the risk assessment, such as the potential for exposure, states must require production of new information. Once the risks are characterized, the risks—not necessarily the underlying data— must be shared with the public.

Judged against this standard, the FracFocus policy fails to meaningfully improve public understanding. The policy is both too broad and too narrow in scope. The site requires well operators to disclose information that is extraneous to risk assessment, such as the purpose of each chemical additive. At the same time, critical data are missing. The site primarily requires disclosure of chemicals that have been identified as hazardous by manufacturers. [379] The data on these chemicals are limited by inadequate reporting and trade secret claims. When the EPA searched FracFocus to find constituents of hydraulic fracturing fluids, for example, the Agency could specifically identify all of the constituents in just 2% of the FracFocus well records. [380] Approximately 15% to 20% of records included at least one chemical that had been designated a trade secret. [381] While a few states require well operators to provide additional data on toxicity or exposure, these data are not on FracFocus. The site also makes it difficult to conduct a nationwide risk assessment because it limits search capabilities and presents the data in individual records rather than in a database. [382] Finally, FracFocus does not communicate the risks to the public, for the obvious reason that these risks have not been calculated.

To satisfy risk science, the GWPC, the IOGCC, and the states must first focus on resolving the uncertainties of experts, not the general public. The FracFocus site should provide experts with a complete database of the chemical compositions of fluids and a repository for other critical risk information. To provide experts with trade secret information, FracFocus may have to limit disclosure to scientists and require confidentiality agreements with those outside government. [383] Additional data that are necessary to risk assessment could be collected by either public or private actors. The oil and gas industry and chemical manufacturers, however, are more likely to do so efficiently. To obtain needed data on toxicity, states could shift the burden to well operators to produce information demonstrating that the hazards of proposed chemical additives are at least as low as those of other additives. [384] To obtain needed data on exposure, states could require operators to conduct testing of nearby freshwater wells and report the results. [385] There must be quality controls to ensure that all of the reported data are useful. Once the risks have been characterized, they should then be communicated to the public.

10.3.2   LEARNING FROM DECISION SCIENCE

From the perspective of decision science, state disclosure policies meaningfully improve public understanding if they ultimately help individuals to make better decisions related to high-volume hydraulic fracturing. In this context, individuals face many different decisions: whether to lease minerals, whether to drink groundwater, whether to buy or sell land, whether to participate in the policy debate. Policies inform these decisions by providing useful data to risk scientists or individuals. In the former case, the policies must provide risk scientists with data that reduce their uncertainty about potential effects of the chemical activity and that also result in knowledge that improves individuals' decisions. In the latter case, the policies must provide individuals with data that, when explained, improve their decisions. States must communicate the information in a neutral way to retain their credibility. The communications must be carefully designed, and yet reach as many individuals as possible before they make decisions. And the final judgment about the acceptability of the risk must be left to individuals.

Judged against this standard, the FracFocus policy fails to meaningfully improve public understanding. For the reasons discussed above, partial data on chemical composition is unlikely to generate expert knowledge, particularly not if the knowledge must inform decisions related to specific wells. The data are also very unlikely to help landowners who live near wells make decisions about their property or interested citizens make decisions about the policy debate—the apparent focus of the site's communications. A landowner who clicks on "Looking for information about a well site near you?" and then opens the well record will not know whether to sell her property after reading the names and concentrations of twenty or more chemical constituents. FracFocus promises to "put th[e] chemical information into perspective" by "provid[ing] objective information." [386] But the information is not connected to specific decisions, [387] and the tone of the communication suggests that individuals should not worry about understanding the chemical data, since the chemicals are present in very small concentrations and stay far underground, and the current laws protect the public from any harm. [388] Even a landowner's decision to test a freshwater well is not informed by the data. [389]

To satisfy decision science, the GWPC, the IOGCC, and the states must focus on resolving the uncertainties that prevent individuals from making good decisions. This requires officials, in partnership with decision scientists, to identify the universe of decisions that individuals must make, learn how individuals understand the chemical activity of high-volume hydraulic fracturing, and compare this understanding to the understanding of experts. Once officials have gathered this information, they can then focus on the decisions. The tone of the materials must be scrupulously neutral to garner public trust, and the site should avoid relying on the conclusions of the oil and gas industry. It should also make clear that its "customers" are members of the general public. [390] One critical aspect of the communication will be conveying the extent of scientific uncertainty in ways that assist individuals in making decisions. This avoids sending the message that no evidence of harmful effects is equivalent to no risk. As more studies are completed, the results should be integrated into the communication.

10.3.3   A PROPOSAL

Combined, the perspectives of risk science and decision science lead to a surprising conclusion: to truly improve public understanding about risk, disclosure of chemicals should be designed to inform experts who can assess that risk. At the same time, individuals must have information so they can make important decisions. To achieve both of these goals, FracFocus should be separated into two sections that respond to the needs of each audience. One section of the site, which may be restricted, should offer risk scientists the most complete data possible on high-volume hydraulic fracturing fluids from across the country. [391] The other section of the site, which must be open to all, should offer individuals the most accurate information they need to make decisions. Utilizing the existing FracFocus site is a pragmatic choice. Disclosure policy has already coalesced around the site. The states have invested significant resources in FracFocus, and the site is well known among stakeholders and the oil and gas industry. But utilizing the site can also be justified normatively. The site has the potential to be a model for cooperative state policy on a controversial environmental and energy issue.

At a minimum, every state should require the operator of each well to disclose the complete identity and concentration of all constituents of fluids after high-volume hydraulic fracturing. To ensure that well operators are held accountable for the information they disclose, the data must be disclosed directly to the state, as part of the state's regulatory program. Operators may request protection for data they consider to be a trade secret, but the states should review the operators' claims. States are then responsible for adding the data to the database on FracFocus so that scientists have access to it. Because of the paucity of exposure data, all states should also require operators to monitor groundwater around wells and report the data to the state for inclusion in the database. In addition to providing important information for risk assessment, this monitoring alleviates the need for individuals to know the chemical composition of fluids in order to conduct baseline testing of freshwater wells.

FracFocus should offer tailored risk communications for each type of decision related to high-volume hydraulic fracturing. Some decisions—such as whether to participate in the policy debate—may require general information about the issue and an explanation of the extent of scientific uncertainty. Other decisions—such as whether to lease minerals—may require both general information on leases and specific information on well operators, such as an operator's compliance history. Yet other decisions—such as whether to buy or sell land—may require very specific information about the location of proposed and existing wells. In addition to narrative descriptions, the site could harness the existing graphical interface to provide needed information. To provide individuals with useful information and to regain public trust, the states should partner with decision scientists trained in the mental models approach.

This proposal is not a panacea. Even with the information on Frac-Focus, members of the public may find it difficult to use the levers of informational regulation to respond to multiple, shortterm sources of risk. The transaction costs of Coasean bargaining are prohibitive when communities must bargain with many operators. Citizen activists may not be able to enforce the terms of a social license against so many potential violators. Participants in democratic deliberation may be overwhelmed by the scale of the chemical activity. And the right to know is diminished when individuals do not know where to focus their attention. [392] Just as there were studies on the TRI, it will be important to continually evaluate the effectiveness of the disclosure policies.

10.4 CONCLUSION

This Article set out to answer the following question: do the state chemical disclosure policies effectively respond to uncertainty about the long-term health and environmental effects of high-volume hydraulic fracturing? In their current form, the unfortunate answer is no. The policies are not likely to reduce either scientific or public uncertainty in a way that meaningfully improves public understanding of risk. In fact, it is more likely that these policies feed public mistrust of state governments and the oil and gas industry. Based on the perspectives of risk science and decision science, this Article proposes a new approach—more Jeffersonian, less Texas Railroad Commission. Rather than "helping the public understand the safety of hydraulic fracturing," [393] this proposal seeks to "inform [the public's] discretion" [394] through better risk assessment and risk communication. If adopted, the new FracFocus-centered policy does not solve the problem of uncertainty. It does, however, require the states to provide much more information to the public on the bounds of that uncertainty. Whether the public will then exercise its control with "a wholesome discretion," [395] we can only see.

ENDNOTES

181. 42 U.S.C. §§ 11001–11050 (2006).
182. 15 U.S.C. §§ 2601–2697 (2006).
183. 42 U.S.C. § 300f (2006).
184. The Federal Insecticide, Fungicide, and Rodenticide Act (FIFRA), 7 U.S.C. §§ 136–136y (2006), governs biocides. Under FIFRA's licensing scheme, the EPA registers a biocide for specific uses, and the product label gives instructions for each use. 7 U.S.C. § 136a (2006). Use that is inconsistent with the label is prohibited. Id. § 136j(a)(2)(G). The EPA has approved glutaraldehyde, a common biocide additive, for use in completion fluids. EPA, EPA 739-R-07-006, REREGISTRATION ELIGIBILITY DECISION FOR GLUTARALDEHYDE 5 (2007). But state pesticide control officials have raised questions about whether other biocides are approved for this use. Gayathri Vaidyanathan, Official Urges EPA Review, Labeling of Fracking Substances, ENERGYWIRE (Oct. 24, 2012), http://www.eenews.net/energywire/2012/10/24/archive/5.

185. Carcinogens that comprise less than 0.1% of the mixture and other constituents that comprise less than 1% of the mixture are not counted towards the reporting thresholds. 40 C.F.R. § 372.38(a) (2012); 29 C.F.R. § 1910.1200(d)(5)(ii) (2013).
186. 42 U.S.C. § 11023(b)(1) (2006); 40 C.F.R. § 372.23 (2012).
187. 42 U.S.C. § 11023(f)(1)(A). If the toxic chemical is manufactured or processed at the facility, the threshold is 25,000 pounds. Id. § 11023(f)(1)(B). The EPA defines the worker threshold as 20,000 hours of work for a facility by the owner, contractors, and employees in a year. 40 C.F.R. § 372.3 (2012).
188. For example, a well pad developed by EOG Resources in Clearfield County, Pennsylvania, contains six wells. There are two reportable chemical substances in the fluid: methanol and propargyl alcohol. Assuming that the fluid has the same density as water, the total amount of methanol used in the wells is 6,671 pounds and the total amount of propargyl alcohol is 766 pounds. See Find a Well, FRACFOCUS, supra note 42 (from the drop-down menus, choose the State of Pennsylvania, the county of Clearfield, and wells "SGL 8H-90," "SGL 13H-90," "SGL 14H-90," "SGL 15H-90," "SGL 16H-90," and "SGL 17H-90").
189. Compare 40 C.F.R. § 372.65 (listing toxic chemicals under the EPCRA), with EPA STUDY PROGRESS REPORT, supra note 75, app. A (listing chemical constituents found in hydraulic fracturing fluid). See also TRI Listed Chemicals, EPA, http://www2.epa.gov/toxics-release-inve ntory-tri-program/tri-listed-chemicals (last visited Oct. 27, 2013) (containing links to lists of reportable chemicals by year).
190. See 42 U.S.C. § 11049(8) (defining a "release" to include pumping or injection).
191. Addition of Facilities in Certain Industry Sectors; Toxic Chemical Release Reporting; Community Right-to-Know, 61 Fed. Reg. 33,588, 33,592 (June 27, 1996) (to be codified at 40 C.F.R. pt. 372). In 2012, environmental organizations petitioned the agency to add the oil and gas extraction industry. Petition to Add the Oil and Gas Extraction Industry, Standard Industrial Classification Code 13, to the List of Facilities Required to Report Under the Toxics Release Inventory (Oct. 24, 2012), available at http://www.foreffectivegov.org/files/in fo/2012.10.24_t ri_petition_final.pdf. The EPA has not yet responded to the petition.
192. A "hazardous chemical," defined in reference to worker safety law, is any chemical that is found to create a health or physical hazard. See 42 U.S.C. §§ 11021(e), 11022(c); 29 C.F.R. § 1910.1200(c).
193. 42 U.S.C. §§ 11021–11022 (2006); 40 C.F.R. § 370.10(a)(2)(i) (2012). The MSDS, a document designed to protect workers handling the product, contains the identity of the hazardous ingredients, the chemical properties, and the known health effects. In March 2012, the MSDS became known as a "Safety Data Sheet," or SDS. 29 C.F.R. § 1910.1200(g) (2013). Because the document is still widely referred to as a MSDS and the new requirements are not effective until 2015, this Article will use the term MSDS.
194. See 40 C.F.R. § 370.10(a)(1) (requiring reporting for extremely hazardous substances if they are present at the facility in amounts greater than 500 pounds or the threshold planning quantity, whichever is less).
195. See infra note 333 and accompanying text.
196. Compare EPA STUDY PROGRESS REPORT, supra note 75, app. A (listing chemical constituents found in hydraulic fracturing fluid, including acrolein, ammonia, an-

iline, benzyl chloride, chlorine, ethylenediamine, formaldehyde, hydrogen fluoride, hydrogen peroxide, methyl vinyl ketone, ozone, peracetic acid, phenol, phosphine, sulfur dioxide, and sulfuric acid), with 40 C.F.R. § 355 app. A (listing the same constituents as extremely hazardous substances).

197. For example, the author requested annual inventories from ten sites with high-volume hydraulically fractured wells in Michigan. Only one well operator had submitted an inventory, and none of the hazardous chemicals appeared to be constituents of hydraulic fracturing fluid. Encana Oil & Gas (USA) Corp., Tier II Emergency and Hazardous Chemical Inventory, State Wilmot 1–21 (2011) (on file with the author) (listing silica, the proppant; diesel, presumably fuel; barium sulfate, an agent to increase the density of drilling muds; and "drilling muds and associated additives"). Of the extremely hazardous chemicals, only one substance has a threshold that appears low enough to require reporting. See 40 C.F.R. pt. 355 app. A (listing a threshold of ten pounds for methyl vinyl ketone).

198. See 42 U.S.C. §§ 11021(a)(3), 11022(a)(3) (2006); 40 C.F.R. § 370.14 (2012).

199. 42 U.S.C. § 11022(e)(3)(C) (allowing the public to request specific Tier II inventory information for a hazardous chemical that is present in an amount less than 10,000 pounds and not in the possession of emergency authorities); 40 C.F.R. § 370.61(3) (same). Emergency authorities have discretion in deciding whether to provide the information, and there are no federal guidelines on how to exercise this discretion. 42 U.S.C. § 11022(e)(3)(C).

200. 29 C.F.R. § 1910.1200(c), (i)(1) (2013) (defining a trade secret as "any confidential formula, pattern, process, device, information or compilation of information that is used in an employer's business, and that gives the employer an opportunity to obtain an advantage over competitors who do not know or use it").

201. See id. § 1910.1200(i)(1)(ii).

202. Id. § 1910.1200(i)(2)–(13).

203. 40 C.F.R. § 350.5(a); Trade Secrecy Claims for Emergency Planning and Community Right-to-Know Information; and Trade Secret Disclosures to Health Professionals, 53 Fed. Reg. 28,772, 28,774 (July 29, 1988) (stating that "[r]egardless of the basis for a trade secret (e.g., a chemical's presence at a facility, its use for a particular process, or its production in a certain quantity), the only information that a facility may withhold from an [EPCRA] report . . . is the specific chemical identity" and in certain circumstances, the location).

204. 40 C.F.R. § 350.5(b)(2)(i) (2012).

205. Id.

206. 42 U.S.C. § 11042(d)(1), (3)(C) (2006).

207. Id. § 11042(h)(1).

208. See 40 C.F.R. § 350.27(b) (stating that a facility is not required to submit an unsanitized version of a MSDS). The operator also need not submit an unsanitized version of an inventory.

209. 15 U.S.C. §§ 2603, 2607 (2006).

210. Id. § 2607(c), (e).

211. See 40 C.F.R. §§ 704.25–.175 (2012) (requiring information only for chemical substances, not mixtures or components of mixtures); see also id. § 716.120 (listing chemical substances and categories of substances for which health and safety studies

must be submitted, but not mixtures). But see id. § 799.5025 (listing some mixtures subject to testing consent orders).

212. 15 U.S.C. § 2603(a) (requiring that the EPA first find that a chemical or mixture "may present an unreasonable risk of injury to health or the environment" or "is or will be produced in substantial quantities," and that the toxicity of a mixture "may not be reasonably and more efficiently determined or predicted by testing the chemical substances which comprise the mixture" before issuing a test rule).

213. The EPA could collect information from service companies if they manufacture a new mixture or "process" the chemical additives into a new mixture. See 15 U.S.C. § 2602(10) (2006) (defining "process" as "the preparation of a chemical substance or mixture, after its manufacture, for distribution in commerce . . . in the same form or physical state as, or in a different form or physical state from, that in which it was received by the person so preparing such substance or mixture, or . . . as part of an article containing the chemical substance or mixture").

214. 15 U.S.C. § 2607(a)(1)(B)(i). In contrast, the EPA could require information on constituents of hydraulic fracturing fluids as long as it acts reasonably. Id. § 2607(a)(1) (A).

215. Id. § 2607(a)(1)(B).

216. See Citizen Petition Under Toxic Substances Control Act Regarding the Chemical Substances and Mixtures Used in Oil and Gas Exploration or Production (Aug. 4, 2011), available at http://www.epa.gov/opt/chemtest/pubs/section_21_Petition_on_ Oil_Gas_Drilling_ and_Fracking_Chemicals8.4.2011.pdf; Letter from Stephen A. Owens, Assistant Adm'r, EPA, to Deborah Goldberg, Earthjustice (Nov. 23, 2011), available at http://www.epa.gov/oppt/chem test/pubs/EPA_Letter_to_Earthjustice_on_TSCA_petition.pdf. The EPA denied the request for a test rule. Letter from Stephen A. Owens, Assistant Adm'r, EPA, to Deborah Goldberg, Earthjustice (Nov. 2, 2011), available at http://www.epa.gov/opt/chemtest/pubs/SO.Earthjusti ce.Response.11.2.pdf.

217. 15 U.S.C. §§ 2607(a)(1)–(2), (c)–(d), 2602(7), 2602(10)–(11) (2006).

218. See id. § 2607(c), (e) (requiring reports of injuries to the health of employees to be kept for thirty years and other reports to be kept for five years).

219. Service companies would be subject to these requirements if they are deemed to "process" chemicals into mixtures for commercial distribution. Id. § 2602(10)–(11); 40 C.F.R. § 717.3(g) (2012); TSCA Section 8(e); Notification of Substantial Risk; Policy Clarification and Reporting Guidance, 68 Fed. Reg. 33,129, 33,137 (June 3, 2003).

220. See 40 C.F.R. §§ 717.3(c), 717.3(i), 717.12(b), 717.12(d) (requiring records of "reactions that may indicate a substantial impairment of normal activities, or long-lasting or irreversible damage to health or the environment," which do not include "known" human health effects or accidental spills or releases to the environment if reported under any federal authority); TSCA Section 8(e); Notification of Substantial Risk; Policy Clarification and Reporting Guidance, 68 Fed. Reg. at 33,138–39 (requiring reports on previously unknown "risk[s] of considerable concern," either because the human health effect is serious or there is "widespread and significant exposure").

221. 40 C.F.R. § 717.10 (requiring records if there is a written or oral "allegation"); TSCA Section 8(e); Notification of Substantial Risk; Policy Clarification and Reporting

Guidance, 68 Fed. Reg. at 33,137 (requiring reports of information that is in the possession of or known by the submitter, which "includes information of which a prudent person similarly situated could reasonably be expected to possess or have knowledge").

222. 15 U.S.C. § 2613(a) (2006) (referencing the exemption in the Freedom of Information Act); 5 U.S.C. § 552(b)(4) (2006) (exempting "trade secrets and commercial or financial information obtained from a person and privileged or confidential" from disclosure).

223. 15 U.S.C. § 2613(c)(1); see also 40 C.F.R. § 711.30 (2012) (listing questions that must be answered when claiming confidentiality).

224. See, e.g., 40 C.F.R. § 711.30 (granting companies the ability to assert a claim of confidentiality for these categories of chemical information under the Chemical Data Reporting rule).

225. 15 U.S.C. § 2613(b).

226. See Claims of Confidentiality of Certain Chemical Identities Contained in Health and Safety Studies and Data from Health and Safety Studies Submitted Under the Toxic Substances Control Act, 75 Fed. Reg. 29,754, 29,756 (May 27, 2010) (stating that chemical identity "has been claimed as confidential in a significant number of health and safety submissions").

227. Id. at 29,756–57.

228. See 42 U.S.C. § 11042(d) (2006) (outlining procedure for public to petition for review of trade secret claims).

229. Id. § 300h(d)(1)(B)(ii) (excluding from the definition of "underground injection" the "injection of fluids or propping agents (other than diesel fuels) pursuant to hydraulic fracturing operations related to oil, gas, or geothermal production activities").

230. Id. § 300h (granting the EPA the authority to promulgate regulations and defining "underground injection" as "subsurface emplacement of fluids by well injection").

231. 40 C.F.R. § 144.6 (2012). The EPA oversees the permitting program in some states, but a majority of states have primacy to enforce the requirements. See UIC Program Primacy, EPA, http://www.water.epa.gov/type/groumfwater/uiv/primacy.cfm (last updated Aug. 1, 2012) (stating that thirty-three states have primacy).

232. 40 C.F.R. § 144.6(b) (including "conventional oil or natural gas production" in Class II wells).

233. Id. § 146.23(b)(1), (c)(1) (requiring an operator to monitor the nature of the injected fluids and report the characteristics annually); id. § 146.24(a)(4)(iii) (requiring an operator to submit the proposed source and an analysis of only the physical and chemical characteristics in the permit application); see also EPA, OMB No. 2040-0042, UNDERGROUND INJECTION CONTROL PERMIT APPLICATION 5 (2011) (listing requirements); EPA, ANNUAL ANALYTICAL REPORT FOR CLASS II INJECTION WELLS, available at http://www.epa.gov/region5/water/uic/forms/annual. pdf (sample disclosure form).

234. See 40 C.F.R. § 146.13(b)(1), (c)(1)(i) (requiring an operator to analyze the characteristics of the actual fluid and report the data to the EPA on a quarterly basis); id. § 146.14(a)(7) (requiring an operator to provide the proposed source of the fluid and an analysis of the chemical, physical, radiological and biological characteristics in the permit application).

235. See EPA, REGION 5–UNDERGROUND INJECTION CONTROL SECTION, RE-
     GIONAL GUIDANCE #8, PREPARING A WASTE ANALYSIS PLAN AT CLASS
     I INJECTION WELL FACILITIES (1994) (referencing the four characteristics of
     ignitability, corrosivity, reactivity, and toxicity in the Resource Conservation and
     Recovery Act); 40 C.F.R. § 261.24 (2012) (describing toxicity testing requirements).
236. EPA, REGIONAL GUIDANCE #8, supra note 235 (applying to the midwestern
     states).
237. See supra notes 36–40 and accompanying text.
238. Permitting Guidance for Oil and Gas Hydraulic Fracturing Activities Using Diesel
     Fuels–Draft: Underground Injection Control Program Guidance #84, 77 Fed. Reg.
     27,451 (proposed May 10, 2012). The guidance would only apply to the EPA, not
     to states that have primacy to implement the program. EPA, PERMITTING GUID-
     ANCE FOR OIL AND GAS HYDRAULIC FRACTURING ACTIVITIES USING
     DIESEL FUELS–DRAFT: UNDERGROUND INJECTION CONTROL PRO-
     GRAM GUIDANCE #84, at 1 (2012).
239. EPA, PERMITTING GUIDANCE FOR OIL AND GAS HYDRAULIC FRAC-
     TURING ACTIVITIES USING DIESEL FUELS, supra note 238, at 20.
240. Id. at 26.
241. See 42 U.S.C. § 300j-4(d)(1) (2006). There is an exception for information that
     "deals with the level of contaminants in drinking water." Id. § 300j-4(d)(2)(B). See
     also 40 C.F.R. § 2.304(e) (2012) (stating that the exception applies to information
     that "deals with the existence, absence, or level of contaminants in drinking water").
242. Permitting Guidance for Oil and Gas Hydraulic Fracturing Activities Using Diesel
     Fuels–Draft, 77 Fed. Reg. 40,354, 40,354 (July 9, 2012).
243. See NAT'L ENERGY TECH. LAB., U.S. DEP'T OF ENERGY, STATE OIL AND
     NATURAL GAS REGULATIONS DESIGNED TO PROTECT WATER RE-
     SOURCES 17–30 (2009), available at http://ene rgyindepth.org/wp-content/up-
     loads/2009/03/oil-and-gas-regulation-report-final-with-cover-5-27 -20091.pdf.
244. Id. at 16.
245. See id. at 18–21.
246. About Us, GROUNDWATER PROT. COUNCIL, http://www.gwpc.org/about-us
     (last visited Oct. 27, 2013).
247. STATE OIL AND NATURAL GAS REGULATIONS DESIGNED TO PROTECT
     WATER RESOURCES, supra note 243, at 19 (stating that 93% of the surveyed
     states require surface casing below the deepest groundwater zone and 96% require
     cement to be circulated from bottom to top).
248. Id. (stating that 78% of the surveyed states require a cement setup period or integrity
     test).
249. Id. at 25.
250. Id.
251. Id.
252. See id.; see also NAT'L ENERGY TECH. LAB., U.S. DEP'T OF ENERGY, STATE
     OIL AND NATURAL GAS REGULATIONS DESIGNED TO PROTECT WATER
     RESOURCES: REGULATIONS REFERENCE DOCUMENT (2009), available
     at http://s3.amazonaws.com/propublica/assets/natu ral_gas/addendum_regs_refer-
     ence_doc.pdf (surveying state regulations).

253. See ALA. ADMIN. CODE r. 400-1-4-.07 (2009) (requiring approval of chemical treatment or fracturing); ALASKA ADMIN. CODE tit. 20, § 25.280 (2010) (requiring approval of stimulation); 55-3 WYO. CODE R. § 22(f) (LexisNexis 2012) (requiring approval of work plans for "stimulation operations" when an operator drills in areas of trona mineral resources); see also LA. ADMIN. CODE tit. 43, § 105(A) (2013) (requiring approval of acidizing and perforation).

254. 55-3 WYO. CODE R. § 22(f)(ii).

255. See N.Y. COMP. CODES R. & REGS. tit. 6, § 554.1(a), (e) (2013) (requiring completion program to prevent pollution and migration of fluids); OKLA. ADMIN. CODE § 165:10-3-10 (2009) (prohibiting fracturing processes from causing pollution).

256. See ARIZ. ADMIN. CODE § 12-7-117 (2007) (requiring immediate notification and diligent correction of damage caused by artificial stimulation); N.M. CODE R. § 19.15.16.17 (LexisNexis 2013) (requiring written notice within five days and diligent correction of damage caused by fracturing a well); N.D. ADMIN. CODE 43-02-03-27 (2010) (requiring diligent correction of damage caused by fracturing).

257. Fracturing Responsibility and Awareness of Chemicals Act, H.R. 2766, 111th Cong. (2009); Fracturing Responsibility and Awareness of Chemicals (FRAC) Act, S. 1215, 111th Cong. (2009).

258. H.R. 2766 § 2(a); S. 1215 § 2(a).

259. H.R. 2766 § 2(b)(1); S. 1215 § 2(b)(1).

260. H.R. 2766 § 2(b)(2); S. 1215 § 2(b)(2).

261. H.R. 2766 § 2(b)(2); S. 1215 § 2(b)(2).

262. See generally NATHAN RICHARDSON ET AL., RES. FOR THE FUTURE, THE STATE OF STATE SHALE GAS REGULATION (2013), available at http://www. rff.org/rff/documents/RFF-Rpt-Stat eofStateRegs_Report.pdf.

263. See id.

264. See MATTHEW MCFEELEY, NATURAL RES. DEFENSE COUNCIL, STATE HYDRAULIC FRACTURING DISCLOSURE RULES AND ENFORCEMENT: A COMPARISON 14 (2012), available at http://www.nrdc.org/energy/files/Fracking-Disclosure-IB.pdf.

265. 55-3 WYO. CODE R. § 45(d) (LexisNexis 2010).

266. 178-00-1 ARK. CODE R. § B-19 (LexisNexis 2013).

267. COLO. CODE REGS. § 404-1:205A (2012).

268. IDAHO ADMIN. CODE r. 20.07.02.055-.056 (2012).

269. IND. CODE § 14-37-3-14.5 (2011) (applying to coalbed methane wells); 20110727 Ind. Reg. 312110432ERA (July 21, 2011) (applying to coalbed methane wells), replaced by 20120725 Ind. Reg. 312120430ERA (July 19, 2012).

270. LA. ADMIN. CODE tit. 43, § 118(C) (2011).

271. MICH. DEP'T ENVTL. QUALITY, SUPERVISOR OF WELLS INSTRUCTION 1-2011, HIGH VOLUME HYDRAULIC FRACTURING WELL COMPLETIONS 3 (2011) (applying to wells that use more than 100,000 gallons of hydraulic fracturing fluid).

272. MONT. ADMIN. R. 36.22.608, .1015, .1016 (2011).

273. N.M. CODE R. § 19.15.16.19 (2012).

274. 25 PA. CODE § 78.122 (2011).

275. TEX. NAT. RES. CODE ANN. § 91.851 (2011); 16 TEX. ADMIN. CODE § 3.29 (2012).

276. W. VA. CODE R. § 35-8-1 to -3 (2011) (emergency rule applying to horizontal wells that require water withdrawals of 210,000 gallons or more in any thirty-day period); W. VA. CODE § 22-6A-7(e) (2011) (applying to horizontal wells that require water withdrawals greater than 210,000 gallons during any thirty-day period).

277. 33 N.Y. Reg. 11 (Sept. 28, 2011) (applying to wells that use more than 300,000 gallons of water for hydraulic fracturing). The first draft is no longer available on the New York Department of Conservation's website.

278. N.D. ADMIN. CODE 43-02-03-27.1 (2012).

279. OHIO REV. CODE ANN. § 1509.10 (2012).

280. OKLA. ADMIN. CODE § 165:10-3-10 (2012) (applying to horizontal wells beginning on Jan. 1, 2013 and other wells on Jan. 1, 2014).

281. UTAH ADMIN. CODE r. 649-3-39 (2013).

282. 58 PA. CONS. STAT. § 3222.1 (2012).

283. IND. CODE § 14-37-3-8 (2012); 20120627 Ind. Reg. 312120292ERA (June 21, 2012).

284. LA. REV. STAT. ANN. § 30:4(L) (2012) (exempting operations "conducted solely for the purposes of sand control or reduction of near wellbore damage").

285. Alaska Oil & Gas Conservation Comm'n, Notice of Proposed Changes in the Regulations (Dec. 20, 2012), http://doa.alaska.gov/ogc/hear/HydraulicFrac.pdf.

286. Cal. Dep't of Conservation, Div. of Oil, Gas & Geothermal Res., Pre-Rulemaking Discussion Draft (Dec. 17, 2012), http://www.conservation.ca.gov/dog/general_information/ Documents/121712DiscussionDraftofHFRegs.pdf.

287. Miss. State Oil & Gas Bd., Statewide Rule 1.26 (June 29, 2012), http://www.sos.ms. gov/ACProposed/00018951b.pdf.

288. Neb. Oil & Gas Conservation Comm'n, Amended Application for an Order Revising the Rules and Regulations of the Comm'n (July 24, 2012), http://www.nogcc. ne.gov/Publicatio ns/AMENDEDNE_Case12-02.pdf.

289. 39 S.D. Reg. 117 (Dec. 17, 2012).

290. Tenn. Dep't of Envtl. Conservation, Rulemaking Hearing Rule(s) Filing Form (Oct. 2012), http://www.tn.gov/environment/wpc/docs/fracking_rules_2012.pdf.

291. 34 N.Y. Reg. 3 (Dec. 12, 2012); see also High Volume Hydraulic Fracturing Proposed Regulations, N.Y. DEP'T ENVTL. CONSERVATION, http://www.dec. ny.gov/regulations/77353. html (last visited Oct. 27, 2013).

292. H.B. 2526 (Kan. 2012) (amending KAN. STAT. ANN. § 55-152(a)).

293. S.B. 820 (N.C. 2012) (amending, among other provisions, N.C. GEN. STAT. § 113-391(a) to require rule for horizontal wells).

294. Oil and Gas; Well Stimulation, Including Hydraulic Fracturing, on Federal and Indian Lands, 77 Fed. Reg. 27,691 (proposed May 11, 2012) (to be codified at 43 C.F.R. pt. 3160).

295. According to the EIA, thirty-three states produced oil or natural gas in 2011. Rankings: Natural Gas Marketed Production, 2011, U.S. ENERGY INFO. ADMIN., http://www.eia.gov/beta/ state/rankings/?sid=US#/series/47 (last visited Oct. 27, 2013) (listing thirty-two states, not including Missouri); Crude Oil Production, U.S. ENERGY INFO. ADMIN., http://www.eia.gov/dn av/pet/pet_crd_crpdn_adc_

mbbl_a.htm (last visited Oct. 27, 2013) (listing thirty-one states, including Missouri). In addition, Idaho and North Carolina anticipate development.

296. CAL. PUB. RES. CODE § 3160 (West 2013).

297. 30 ILL. COMP. STAT. 105/5.826 (2013).

298. 26-2 MISS. CODE R. § 1.26 (LexisNexis 2013).

299. TENN. COMP. R. & REGS. 0400-53-01-.03 (2013).

300. Nev. Comm'n on Mineral Res., Resolution Concerning Hydraulic Fracturing (Feb. 22, 2013), http://minerals.state.nv.us/forms/ogg/COMMISSION%20ON%20MINERAL%20RESOURCES%20RESOLUTION.pdf

301. S.D. ADMIN. R. 74:12:02:19 (2013).

302. W. VA. CODE R. § 35-8-5.6.b.5, -10.1 (2013).

303. Alaska Oil & Gas Conservation Comm'n, Notice of Proposed Changes in the Regulations (Aug. 7, 2013), http://doa.alaska.gov/ogc/frac/02_01_Hydraulic%20Fracturing%20Public%20N otice%20and%20Additional%20Info.pdf; Alaska Oil & Gas Conservation Comm'n, Notice of Proposed Changes in the Regulations (Nov. 1, 2013), http://doa.alaska.gov/ogc/frac/03_01_Hyd raulic%20Fracturing%20Public%20Notice%20and%20Additional%20Info.pdf.

304. Neb. Oil & Gas Conservation Comm'n, Second Amended Application for an Order Revising the Rules and Regulations of the Comm'n (Apr. 3, 2013), http://www.nogcc.ne.gov/ Publications/NE_RulesRegsChangesAmend2.pdf.

305. Notice of Hearing on Proposed Administrative Regulations, 32 Kan. Reg. 543 (May 30, 2013).

306. Mich. Dep't of Envtl. Quality, Oil and Gas Operations, Proposed Draft (Nov. 1, 2013), http://www.michigan.gov/documents/deq/DRAFT_Hydraulic_Fracturing_Rules_438152_7.pdf.

307. N.C. Mining & Energy Comm'n, Final ESC Draft to Rules Committee (Apr. 1, 2013), http://portal.ncdenr.org/c/document_library/get_file?uuid=e6d7d88a-73a0-49d4-8404-9ad6d3c 790f4&groupId=8198095.

308. MD. DEP'T OF THE ENV'T & MD. DEP'T OF NATURAL RES., MARCELLUS SHALE SAFE DRILLING INITIATIVE STUDY PART II: BEST PRACTICES 28–29 (2013).

309. H.B. 743, 2013 Leg., Reg. Sess. (Fl. 2013) (requiring disclosure, which passed in House but died in Senate committee); H.B. 745, 2013 Leg., Reg. Sess. (Fl. 2013) (specifying trade secret treatment, which died on House calendar).

310. Ariz. Oil & Gas Conservation Comm'n, Minutes of Meeting (Apr. 19, 2013), http:// azog cc.az.gov/sites/azogcc.az.gov/files/meetings/M2013.04.19_0.pdf.

311. Oil and Gas; Hydraulic Fracturing on Federal and Indian Lands, 78 Fed. Reg. 31,636 (proposed May 24, 2013) (to be codified at 43 C.F.R. pt. 3160).

312. In 2000, Alabama adopted rules on coalbed methane wells in response to litigation over the applicability of the SDWA to the state's permitting program. Under these rules, the operator of certain coalbed methane wells must submit "the type [of] fluids and materials that are to be utilized" to the state as part of a proposed fracturing program. ALA. ADMIN. CODE R. 400-3-8-.03(5) (2013). This requirement lacks, however, the specific focus on chemical constituents found in recent disclosure policies.

313. See, e.g., Mead Gruver, Wyoming Approves Fracking Disclosure Rules, BOSTON. COM (June 8, 2010) (quoting Wyoming Governor Freudenthal as saying that it is

imperative that hydraulic fracturing continue, "[b]ut it is imperative that it continue in a way that is properly supervised and overseen by the Wyoming Oil and Gas Commission").

314. Mike Soraghan, Natural Gas Company's Disclosure Decision Could Change Fracking Debate, GREENWIRE (July 15, 2010) (stating that disclosure "reflects the desire of industry to get out ahead of the issue to prevent federal regulation of the key drilling practice called hydraulic fracturing").

315. See, e.g., About Us, FRACFOCUS, http://fracfocus.org/welcome (last visited Oct. 27, 2013) (stating that the "primary purpose of this site is to provide factual information concerning hydraulic fracturing and groundwater protection").

316. Cf. FUNG ET AL., supra note 115, at 39–49 (describing five common characteristics of "targeted transparency" policies).

317. As of 2009, twenty-two states had adopted a database system known as the Risk-Based Data Management System. See Interstate Oil & Gas Compact Comm'n, Risk-Based Data Management System, GROUNDWORK, May 11, 2009, at 3. The system was originally designed to provide states with data on injection wells, and then was expanded to include production wells and an "e-commerce" platform for the oil and gas industry. Id. at 4. Some states also give the public access to certain database information. See, e.g., infra note 331.

318. 25 PA. CODE § 78.122(b)(6) (2011).

319. Id. § 78.122(b)(6)(i)–(ii).

320. Id. § 78.122(b)(6)(iii). The CAS number is the "gold standard" of chemical identification. See CAS REGISTRY–The Gold Standard for Chemical Substance Information, CHEMICAL ABSTRACTS SERVICE, http://www.cas.org/content/chemical-substances (last visited Oct. 27, 2013).

321. 25 PA. CODE § 78.122(b)(6)(iv).

322. Id. § 78.122(d).

323. See generally 65 PA. CONS. STAT. §§ 67.101–.3103 (2012) (Pennsylvania's Right-toKnow Law).

324. 25 PA. CODE § 78.122(c).

325. 55-3 WYO. CODE R. § 45(d) (LexisNexis 2011) (emphasis in original omitted).

326. Id. § 45(d)(i), (iii).

327. Id. § 45(d)(ii).

328. Id. § 45(g). The Wyoming Oil and Gas Conservation Commission clarified that it would focus on the listed substances and not on the entire class of volatile organic compounds. Memorandum from Thomas E. Doll, State Oil & Gas Supervisor, to All Operators (Aug. 24, 2010), http://groundwork.iogcc.org/sites/default/files/WY_Memo_adopt_Rules_Aug2010.pdf.

329. 55-3 WYO. CODE R. § 45(h)(ii). Although this part of the rule only refers to chemical additives, the Wyoming Oil and Gas Conservation Commission expects operators to report the information by chemical constituent. See, e.g., Nicholas Kusnetz, Wyoming Fracking Rules Would Disclose Drilling Chemicals, PROPUBLICA (Sept. 14, 2010), http://www.propublica.org/ article/wyoming-fracking-rules-would-disclose-drilling-chemicals (quoting the Supervisor of the Commission as saying, "What we've explained to the operators and what we expect is each of these components, whatever is in that mix, will have to be disclosed.").

330. 55-3 WYO. CODE R. § 45(f); WYO. STAT. ANN. § 16-4-203(d)(v) (2011) (exempting "[t]rade secrets, privileged information and confidential commercial, financial, geological or geophysical data furnished by or obtained from any person" from disclosure).

331. See WYO. OIL & GAS CONSERVATION COMM'N, http://wogcc.state.wy.us/ (last visited Oct. 27, 2013) (follow "Wells" hyperlink to search database by number, well name, and location, or follow "Completions" hyperlink to search by date, then select a well, click on + sign next to "Approvals/Notice," click on + sign next to "Sundries Form 4," then follow "Notice of Intent Fracture Treat/Enhance" hyperlink or "Subsequent Report Fracture Treat/Enhance" hyperlink to view forms; for trade secret submissions and the Commission's responses, begin at the main page and follow the "Notices, Memo's [sic] and Details" hyperlink, and then the "Approved Trade Secrets" hyperlink).

332. 16 TEX. ADMIN. CODE § 3.29(c)(1) (2012).

333. Id. §§ 3.16(b), 3.29(a)(8), 3.29(c)(2)(A); FracFocus Chemical Disclosure Registry, FRACFOCUS, http://fracfocus.org/ (last visited Oct. 1, 2013).

334. About Us, FRACFOCUS, supra note 315; About Us, INTERSTATE OIL & GAS COMPACT COMM'N, http://www.iogcc.state.ok.us/about-us (last visited Nov. 4, 2013).

335. 16 TEX. ADMIN. CODE § 3.29(c)(2)(A)(ix).

336. Id. § 3.29(c)(2)(A)(x)–(xiii).

337. Id. § 3.29(c)(2)(B).

338. See Find a Well, FRACFOCUS, supra note 42 (showing each information sheet, which includes not just the chemical information, but the date of hydraulic fracturing, the type of well, the vertical depth, and the total volume of water used).

339. 16 TEX. ADMIN. CODE § 3.29(c)(4), (e)–(g).

340. Id. § 3.29(c)(2)(C).

341. COLO. CODE REGS. § 404-1:205A(B)(2) (2012) (requiring disclosure to FracFocus unless inoperable).

342. LA. ADMIN. CODE tit. 43, § 118(C)(4) (2011) (allowing disclosure to FracFocus).

343. Nev. Comm'n on Mineral Res., supra note 300 (requiring disclosure to FracFocus).

344. N.D. ADMIN. CODE 43-02-03-27.1(g) (2012) (requiring disclosure to FracFocus).

345. OHIO REV. CODE ANN. § 1509.10(F) (West 2012) (allowing disclosure to FracFocus).

346. OKLA. ADMIN. CODE § 165:10-3-10(b) (2012) (allowing disclosure to FracFocus or to the state, in which case the state will post the information on FracFocus).

347. S.D. ADMIN. R. 74:12:02:19 (2013) (requiring disclosure to FracFocus).

348. 16 TEX. ADMIN. CODE § 3.29(a)(8), (c)(2)(A) (2012) (requiring disclosure to FracFocus unless inoperable).

349. UTAH ADMIN. CODE r. 649-3-39(1) (2013) (requiring disclosure to FracFocus).

350. Kan. Corp. Comm'n, Notice of Hearing on Proposed Administrative Regulations (June 4, 2013), http://kcc.ks.gov/pi/hearing_kar_081513.htm (proposing to create KAN. STAT. ANN. §§ 82-3-1400 to -1402, which would require disclosure to FracFocus or state).

351. Mich. Dep't of Envtl. Quality, supra note 306 (proposing to create MICH. ADMIN. CODE r. 324.1406, which would require disclosure to FracFocus unless unavailable).

352. Neb. Oil & Gas Conservation Comm'n, supra note 288 (proposing to create 267 NEB. ADMIN. CODE § 3-43, which would require disclosure to FracFocus).

353. N.C. Mining & Energy Comm'n, supra note 307 (proposing disclosure to FracFocus unless inoperable and advance disclosure of chemicals by "approved contractors").

354. CAL. PUB. RES. CODE § 3160(g) (West 2013) (allowing disclosure to FracFocus while state creates its own website).

355. 26-2 MISS. CODE R. § 1.26(11) (2013) (allowing disclosure to FracFocus).

356. MONT. ADMIN. R. 36.22.608, .1010, .1015, .1016 (2011) (allowing disclosure to FracFocus).

357. 58 PA. CONS. STAT. § 3222.1(b)(1)–(2) (2012) (requiring disclosure to FracFocus).

358. TENN. COMP. R. & REGS. 0400-53-01-.03 (2013) (requiring disclosure to Frac-Focus).

359. W. VA. CODE R. §§ 35-8-5.6.b.5, -10.1 (requiring operators of horizontal wells to disclose to FracFocus).

360. Alaska Oil & Gas Conservation Comm'n, supra note 285 (proposing to create ALASKA ADMIN. CODE tit. 20, § 25.283(a)(13), (h)–(i), which would require disclosure to both FracFocus and the state).

361. N.Y. Dep't Envtl. Conservation, Revised Express Terms 6 NYCRR Parts 550 through 556 and 560 (2012), http://www.dec.ny.gov/docs/administration_pdf/rhvh-fet550556570.pdf (proposing to create N.Y. COMP. CODES R. & REGS. tit. x, § 560.3(d), .5(h), which would require disclosure to both FracFocus and the state).

362. Arkansas requires service companies to prepare a master list of all chemicals used in hydraulic fracturing treatments in the state. 178-00-1 ARK. CODE R. § B-19(l) (3) (LexisNexis 2013). The state Oil and Gas Commission then posts the lists on its website. See Well Fracture Information, ARK. OIL & GAS COMM'N, http://www.aogc.state.ar.us/Well_Fracture_Companies.htm (last visited Nov. 7, 2013).

363. 30 ILL. COMP. STAT. 105/5.826 § 1-110 (2013) (directing state to "create and maintain an online searchable database that provides information related to high volume horizontal hydraulic fracturing operations on wells that, at a minimum, include . . . chemical disclosure information").

364. See N.D. ADMIN. CODE 43-02-03-27.1(g) (2012) (requiring operator to "post on the fracfocus chemical disclosure registry all elements made viewable by the frac-focus website"); 58 PA. CONS. STAT. § 3222.1 (2012) (requiring operator to "complete the [FracFocus] form and post the form on [FracFocus] . . . in a format that does not link chemicals to their respective hydraulic fracturing additive"); UTAH ADMIN. CODE r. 649-3- 39(1) (2012) (requiring operator to report the "amount and type of chemicals used in a hydraulic fracturing operation . . . to www.fracfocus.org").

365. COLO. CODE REGS. § 404-1:205A(b)(3) (2012) (requiring disclosure through a state website if FracFocus is not made searchable by geographic area, ingredient, CAS number, time period, and operator); 58 PA. CONS. STAT. § 3222.1(b) (6) (2012) (requiring investigation into the feasibility of disclosure through a state website if FracFocus is not made searchable by the same criteria).

366. Stan Belieu, Deputy Dir., Neb. Oil & Gas Conservation Comm'n, Presentation on FracFocus Chemical Disclosure Registry at the 19th Annual IPEC Conference (Oct. 30, 2012).

367. See, e.g., Find a Well, FRACFOCUS, supra note 42 (as of November 5, 2013, showing 52 wells in Alaska and 315 wells in Kansas, both states that have only proposed a disclosure policy).
368. See, e.g., id. (from the drop-down menu, choose the state of Oklahoma, then click on any well marker for the information sheet, which shows the maximum concentration in the additive even though the state does not require the information).
369. See, e.g., id. (as of November 5, 2013, showing 2,072 wells in Wyoming and 1,605 wells in New Mexico, both of which require reporting only to the state).
370. See generally KATE KONSCHNIK ET AL., ENVTL. LAW PROGRAM POLICY INITIATIVE, HARVARD LAW SCH., LEGAL FRACTURES IN CHEMICAL DISCLOSURE LAWS: WHY THE VOLUNTARY CHEMICAL DISCLOSURE REGISTRY FAILS AS A REGULATORY COMPLIANCE TOOL (2013), available at http://blogs.law.harvard.edu/environmentallawprogram/files/2013/04/4-23-2013-LEGAL-FRACTURES.pdf (arguing that FracFocus is "not an acceptable regulatory compliance method for chemical disclosure" due to three general shortcomings: the timing of disclosure, the substance of disclosure, and the extent of nondisclosure).
371. OHIO REV. CODE ANN. § 1509.10(A)(10)(b), (F) (2012).
372. See Find a Well, FRACFOCUS, supra note 42 (from the drop-down menu, choose the state of Ohio, then click on any well marker for the information sheet).
373. See, e.g., COLO. CODE REGS. § 404-1:205A (2012).
374. About Us, FRACFOCUS, supra note 315 ("The listing of a chemical as proprietary on the fracturing record is based on the 'Trade Secret' provisions related to Material Safety Data Sheets (MSDS) . . . at 1910.1200(i)(1).").
375. Compare 16 TEX. ADMIN. CODE § 3.29(c)(2)(C), with Find a Well, FRACFOCUS, supra note 42 (from the drop-down menu, choose the state of Texas, then click on any well marker for the information sheet). But see Stan Belieu, Debunking Myths about FracFocus, GROUNDWATER COMMUNIQUE NEWSLETTER (Groundwater Prot. Council), Oct. 2012, at 1 (stating that the website discloses "ALL of the information required by the[ ] states with respect to 'hydraulic fracturing chemical disclosure' ").
376. FracFocus Chemical Disclosure Registry, FRACFOCUS, supra note 333.
377. Compare KONSCHNIK ET AL., supra note 370, at 1, with FracFocus Responds to Harvard Study, FRACFOCUS, http://fracfocus.org/node/344 (Apr. 24, 2013) ("FracFocus not only notifies states of the submission of disclosures and provides them with lists of such disclosures on a routine basis, it allows states to download the data from the disclosures so that it can be incorporated into the states [sic] own data system.").
378. CAL. PUB. RES. CODE § 3160(j) (West 2013); see also MCFEELEY, supra note 264, tbl.V (displaying table showing that Colorado, Ohio, and Pennsylvania require the operator to submit a written submission to the state).
379. Frequently Asked Questions, FRACFOCUS, http://fracfocus.org/faq (last visited Oct. 27, 2013) (follow "What chemicals are being disclosed on this website?" hyperlink for answer, "All chemicals that would appear on a Material Safety Data Sheet (MSDS) that are used to hydraulically fracture a well . . . "). When states mandate

disclosure of all constituents, the FracFocus form allows well operators to list the non-hazardous constituents separately from the hazardous constituents in a product.

380. EPA STUDY PROGRESS REPORT, supra note 75, at 61 (stating that 184 of 12,173 well records "had ingredient lists that fully matched the EPA CASRN list").

381. Id. ("Operators reported at least one chemical ingredient as 'CBI,' 'proprietary,' or 'trade secret' in 1,924 well records."); see also Ben Elgin et al., Fracking Secrets by Thousands Keep U.S. Clueless on Wells, BLOOMBERG (Nov. 30, 2012, 12:01 AM), http://www.bloomberg.com/ne ws/2012-11-30/frack-secrets-by-thousands-keep-u-s-clueless-on-wells.html (reporting that "[d]rilling companies in Texas . . . claimed [trade secret] exemptions about 19,000 times" between January and August of 2012, and "[n]ationwide, companies withheld one out of every five chemicals they used in fracking").

382. See Find a Well, FRACFOCUS, supra note 42. Cf. Mike Soraghan, Hydraulic Fracturing: FracFocus Can't Replace Full, Public Disclosure, Groups Say, ENERGY-WIRE (May 21, 2012), http://eenews.net/public/energywire/2012/05/21/1 (quoting Mike Paque, Executive Director of the GPC, as saying "[w]e did not set out to build a national environmental analytic tool or website . . . I guess no good deed goes unpunished"). An organization called SkyTruth "scraped" data from FracFocus to create a database for 2011 and 2012 wells. Fracking Chemical Database, SKYTRUTH, http://frack.skytruth.org/fracking-chemical-database (last visited Oct. 27, 2013).

383. Cf. Letter from Thomas Field et al. to Cathy P. Foerster, Comm'r, Alaska Oil & Gas Conservation Comm'n (Apr. 1, 2013) (letter by several law professors arguing that Alaska should not provide protection for asserted trade secrets).

384. See N.Y. DEP'T ENVTL. CONSERVATION, supra note 361 (proposing to create N.Y. COMP. CODES R. & REGS. tit. x, § 560.3(d)(1)(viii), which would require operators to demonstrate that the proposed additives "exhibit reduced aquatic toxicity" and "pose at least as low a potential risk to water resources and the environment as all known available alternatives," or show that the alternatives are not as effective or are not economically feasible).

385. See COLO. CODE REGS. § 404-1:609 (2013) (requiring baseline testing before hydraulic fracturing, within six to twelve months after, and again within five to six years); 30 ILL. COMP. STAT. 105/5.826 § 1-80(b) (2013); N.Y. Dep't Envtl. Conservation, supra note 361 (proposing to create N.Y. COMP. CODES R. & REGS. tit. x, § 560.5(d)).

386. About Us, FRACFOCUS, supra note 315.

387. See, e.g., What Chemicals Are Used, FRACFOCUS, http://fracfocus.org/chemical-use/wh at-chemicals-are-used (last visited Oct. 27, 2013) (listing the most commonly used chemicals, offering tips on how to learn the specific identity of a chemical, and providing links to federal government and university websites that contain more information on health and environmental effects of individual chemicals).

388. See Chemical Use in Hydraulic Fracturing, FRACFOCUS, http://fracfocus.org/water-pro tection/drilling-usage (last visited Oct. 27, 2013) (stating that "chemicals are needed to insure that the fracturing job is effective and efficient," and noting that a "typical fracture treatment will use very low concentrations of between 3 and 12 additive chemicals"); Groundwater Protection & Water Usage, FRACFOCUS, http://fracfocus.org/groundwater-prot ection (last visited Oct. 27, 2013) ("Pure, clean

groundwater. Nothing can replace it. That is why fresh-water aquifers are protected through strictly regulated exploration and production practices."); Hydraulic Fracturing: The Process, FRACFOCUS, http://fracfocus.org/ hydraulic-fracturing-how-it-works/hydraulic-fracturing-process (last visited Oct. 27, 2013) (presenting text and figures from a 2010 magazine article by an employee of Pinnacle, a Halliburton Company, including the author's conclusion that "hydraulic fractures are not growing into groundwater supplies, and therefore, cannot contaminate them").

389. Groundwater Quality and Testing, FRACFOCUS, http://fracfocus.org/groundwater-prote ction/groundwater-quality-testing (last visited Oct. 27, 2013) (advising nearby landowners to test their wells before oil or gas wells are hydraulically fractured, and to consult with professionals after the disclosure so they can decide whether to conduct additional testing).

390. Compare FracFocus Chemical Disclosure Registry, FRACFOCUS, supra note 333 (stating that upgrades will "dramatically enhance the site's functionality for the public, state regulatory agencies and industry users"), with Press Release, FracFocus, FracFocus 2.0 to Revolutionize Hydraulic Fracturing Chemical Reporting Nationwide (May 29, 2013), http://fra cfocus.org/node/347 (stating that upgrades will "dramatically improve the site's functionality for state regulatory agencies, industry and public users"). See also Benjamin Haas et al., Fracking Hazards Obscured in Failure to Disclose Wells, BLOOMBERG (Aug. 14, 2012, 6:26 PM), http://www.bloomberg.com/news/2012-08-14/fracking-hazards-obscured-in-failure-to-disc lose-wells.html (reporting that the American Petroleum Institute and America's Natural Gas Alliance pay part of the site's operating costs).

391. Almost all new oil or gas wells are hydraulically fractured. Because the chemical activity of concern is high-volume hydraulic fracturing, the database should focus on these types of wells. States will need to agree on the definition of "high-volume" hydraulic fracturing.

181. This is not to say that the prospect of disclosure has had no effect. Two service companies announced "green" chemical additives after the FRAC Act was introduced in Congress, although no toxicity data have been released. See, e.g., Press Release, Halliburton, Halliburton Introduces CleanStim™ Fracture Formulation, Launches New Microsite on Hydraulic Fracturing Fluids Disclosure (Nov. 15, 2010), http:// halliburton.com/public/news/pu bsdata/press-release/2010/corpnws_111510.html; Green Frac, CHESAPEAKE ENERGY, http:// www.chk.com/Corporate-Responsibility/EHS/Environment/Green-Frac/Pages/Information. aspx (last visited Oct. 27, 2013).

182. See Press Release, R.R. Comm'n of Texas, supra note 2.

183. See Letter from Thomas Jefferson to William Charles Jarvis, supra note 1.

184. See id.

CHAPTER 11

Hydraulic Fracturing: Paving the Way for a Sustainable Future?

JIANGANG CHEN, MOHAMMED H. AL-WADEI,
REBEKAH C. M. KENNEDY, AND PAUL D. TERRY

11.1 INTRODUCTION

The United States struggles with increasing carbon emissions due to the use of high-carbon energy sources such as petroleum and coal, which together provide the largest portion of primary energy consumption in the country [1]. Energy-related activities have been the primary source of domestic anthropogenic greenhouse gas (GHG) emissions which contributes to the widespread climate-related stress on water resources, livestock, ecosystems, and human health [2]. This appreciation therefore highlights the link between our primary future energy source(s) and future climate change and impacts.

Solar, wind, biomass waste, and geothermal and hydroelectric energy have long been recognized as renewable and sustainable energy resources; currently however, they only comprise 9% of our energy consumption;

this is in sharp contrast to the rapid growth of national natural gas market production with a record high of 25,319 billion cubic feet (717 billion cubic meters) in 2012 [3]. In fact, natural gas contributed approximately 27% of the total United States energy consumption and accounted for 40% of industrial and 74% of commercial and residential energy consumption in 2012 [1, 3]. Although the accuracy of GHG emission estimates from natural gas production and usage is still a matter of debate [4, 5], natural gas, which is composed mainly of methane, is considered cleaner-burning than coal or oil with significantly lower levels of carbon dioxide, nitrogen oxides (NOx), sulfur dioxide, and particles emission when combusted.

The United States is the largest natural gas producer in the world [6]. Total marketed production grew by 7.9 percent in 2011, which was the sixth consecutive year of growth in marketed production and the largest year-over-year percentage increase since 1984. The increase in natural gas production in United States came exclusively from the onshore which was largely concentrated in shale plays [7]. Largely located in western, midwestern, and northeastern areas of the country, shale gas accounts for about 25% of the total domestic natural gas production with its contribution is still rapidly growing [8]. The utilization of shale gas as a resource would not be considered practical and commercially profitable without the emergence of hydraulic fracturing.

Hydraulic fracturing is an advanced stimulation technological process accompanied by increasing lateral horizontal well drilling (more than 1 km or 3000 ft) and the injection of fracturing fluid under high pressure (480–850 bar) to open new or enlarge the existing rock fractures that facilitate the migration of natural gas toward the surface. Hydraulic fracturing could create a contact area that is thousands of times greater than that achieved by the typical method of vertical drilling therefore significantly increases the production of natural gas from a single well site [9].

Hydraulic fracturing fluid is essential for creating fractures. Fracturing is conducted in multiple stages, and wells may be refractured multiple times to maximize its economic life [8]. Based on formation characteristics, different combinations of fracturing fluids might be used to enhance fracturing effectiveness. While oil, water, methanol, or a combination of water and methanol are used as fracturing fluids in practice, the predominant types of fracturing fluids for unconventional shale gas extractions are water based and this review therefore will focus on the overall impact of water based hydraulic fracturing activities on surface and groundwater.

The fracturing fluid is comprised of approximately 90% water to effectively create thin, long fractures distributed over a larger area, 9% sand (natural occurring sand grains, resin coated sand, high-strength ceramic materials, and resin coated ceramic materials) which serves as proppant to hold or prop open the fractures, and a small percentage of additives serving various purposes [10]. Not all of the additives are used in every hydraulically fractured well (see [11] for a list of additives used in hydraulic fracturing). The composition and proportion of fracturing fluid chemistry is designed according to the geological characteristics of each target formation. As part of the fracturing fluid, acids are used to clean wellbore and dissolve near-wellbore acid soluble minerals (i.e., Calcite $CaCO_3$) to create open conduit for the hydraulic fracturing fluid. For example, hydrochloric acid is one of the most commonly used acids with a typical concentration range of 0.08%–2.1% of the total fluid pumped (as volume of 15% HCl) [12, 13]. Due to the corrosive nature of acids, a corrosion inhibitor at a concentration of 2,000 to 5,000 ppm which equates to 0.0004%–0.0043% of the total fluid volume is added into the fracturing fluid. The acid inhibitor bonds to the surface of metal minimizing the galvanic corrosion of steel tubes, well casing, tools, and fracturing fluid containment tanks, to prevent potential chemical leakage [12, 13]. Higher concentrations of the corrosion inhibitor are required when casing and tubing have a higher composition of metal or when downhole temperatures exceed 250 degrees Fahrenheit (121°C) [13, 14].

To function properly, fracturing fluids need to be viscous enough to create a fracture of adequate width while at the same time be able to travel further away from the well pad to extend fracture length before settling [15]. To achieve this, a more viscous fluid, lighter proppant, or a combination can be applied. There are usually two ways to increase a fluid viscosity. One way is to add a gelling agent, a polymer such as guar or guar derivatives. However, in the presence of high temperatures, the gelled fluid could lose its viscosity, a problem that could be resolved by increasing the polymer concentration or adding cross-linking agents to increase the molecular weight of the solution [14–16]. While the addition of these agents could further increase viscosities by several orders of magnitude, these relative large agents can plug the small pores of the fracture surface and decrease the gas flow [15]. Alternatively, a foam-based fracturing fluid could be used which gives an effective viscosity similar to that of a gelled fluid. The addition of foam in fracturing fluid has the advantage of minimizing the water use, reducing the leaking-off rate, and improving fluid recovery efficiency, although the tradeoffs could be high cost, high surface pump pressure, and low proppant load [15].

Practically speaking, fracturing fluids also include potassium chloride, breakers, biocides, fluid-loss additives, and friction reducers [12, 14, 15]. While most companies rely on the use of gel additives to increase fracture fluid viscosity, in practice, others also add small amounts of potassium chloride to further enhance fracturing fluids viscosity, a step that could reduce the amount of gel required to achieve the same level of viscosity [15]. This approach is considered by some companies to be environmentally "responsible" or "safe" because the current environmental impact of gels is still largely unknown and potassium chloride is a nonapparent health hazard at low concentration [10, 12, 16]. In addition to increasing the fracturing fluid viscosity, 1–3% potassium chloride solutions have been applied in formations to stabilize the clay and prevent its swelling due to the presence of water [17].

Biocides, such as glutaraldehyde, are added to eliminate bacterial growth in fracturing fluids [12, 14, 16]. The growth of bacteria in the presence of organic materials in fracturing fluid can produce corrosive byproducts and enzymes that interfere with gel formation and thereby reduce fluid viscosity [17]. Friction reducers, which include latex polymers or copolymers of acrylamides, are generally mixed with the fracturing fluid at a concentration range of 0.25 to 2.0 pounds per 1,000 gallons (0.11 to 0.90 kg per 3,785 liters) of fluid to minimize the loss of pressure due to the presence of friction between fracturing fluid and tubing/wellbore during the fracturing [12, 16]. Breakers, such as ammonium persulfate, are used in fracturing fluid, to allow a delayed breakdown of the "crosslinker" and "gel" in the formation to reduce fracturing fluid viscosity thereby enhancing postfracturing fluid recovery and flowback [12, 16].

Currently, there are no federal disclosure standards mandating hydraulic fracturing companies to disclose a list of their toxic chemicals [18]. As shown in Table 1, of the 29 states with hydraulic fracturing activity ongoing, only 15 states enact disclosure laws. Of these 15 states, nine states are exempted from disclosing chemicals that serve as trade secrets, and only one of these 9 states (Wyoming) has a process in place for the state evaluation of trade secrets with a factual justification required. Six of the 15 states, however, require the disclosure of trade secret chemicals to health care professionals under emergency circumstances for effective patient treatments but four of the six states further require physicians to sign confidentiality contracts to receive such disclosure which potentially may slow the flow of information during emergencies [18]. Without full disclosure of all chemicals used in fracturing process, it will be difficult to assess the health and environmental impacts [19, 20].

TABLE 1: Chemical disclosure requirements by states.

Chemical disclosure required to the state		No chemical disclosure required to the state
Disclosure of trade secret to medical personnel	No disclosure of trade secret to medical personnel	
Arkansas	Alabama	Alaska, California
Colorado*	Indiana	Illinois, Kansas
Montana*	Louisiana	Kentucky, Mississippi
Ohio	Michigan	Missouri, Nebraska
Pennsylvania*	New Mexico	New York, South Dakota
Texas*	North Dakota	Tennessee, Utah
	Oklahoma	Virginia, Washington
	West Virginia	
	Wyoming	

States requiring physicians to sign confidentiality agreements.

11.2 IMPACT OF HYDRAULIC FRACTURING ON WATER

Before the emergence of new advanced technology, hydraulic fracturing will continuously play an essential role to facilitate the expansion of natural gas development [21]. Two decades ago, available scientific research was focused more on how to improve the efficiency of hydraulic fracturing performance rather than its environmental impacts. Hydraulic fracturing is performed in several stages of varying distance to stimulate the entire length of the well [22]. Each stage requires tens of thousands of barrels of water which can total up to several million gallons per well [22]. The scarcity of peer-reviewed data addressing any association between the technology and the availability and quality of local water resources has been identified as the foremost issue among other social, economic, and environmental concerns [23].

11.2.1 LARGE VOLUME OF WATER WITHDRAW FOR FRACTURING ACTIVITY

Hydraulic fracturing requires the use of large volumes of water for a single operation. Fracturing shale gas typically requires the withdrawal of 2.3–3.8 million gallons (8.7–14.4 million liters) of water per single well [23, 24]. Recent data, however, indicate that the volume of water used during hydraulic fracturing may be underestimated [25]. The amount of water required for fracturing treatments therefore could vary depending on the type of well drilled and its geological location. In general, the deeper the well is and the stronger the rock formation is, the more water is needed for the fracturing process [21, 26].

Industry argues that water used in the hydraulic fracturing process is insignificant compared to the total annual water withdrawal in the United States. In 2005, approximately 149,650 billion gallons (1,552 billion liters per day) of water was withdrawn for various uses in the United States [27]. The largest two sectors for water withdrawal in 2005 were used for thermoelectric power generation (49%) and irrigation (31%) [27]. In contrast, less than 1.0% of water withdrawal was used for mining purposes which includes water used for extracting solid minerals, such as copper; liquids, such as petroleum; and gases, such as natural gas [27]. Even though the booming of hydraulic fracturing process across the nation in most recent years could imply more water withdrawal compared to 2005, the portion of the water withdrawal overall used for natural gas production after 2005 should not change significantly [27].

Nonetheless, even 2 million gallons (7.6 million liters) of water used per well could be significant, simply because the water is usually taken directly from one single location and in many cases from remote and environmentally sensitive areas. While total hydraulic fracturing water use represents less than 1.0% of the water use in the nation [27], the hydraulic fracturing water use is unevenly distributed across individual states and may locally represent a higher fraction of the total water use which can result in a significant impact on local flow regime [28, 29]. Texas for instance, is an area with a wet and dry season. Total water use for hydraulic fracturing (for oil and natural gas) in Texas has increased by about 125%, from 36,000 acre feet (AF) (0.04 cubic kilometers) in 2008 to about 81,500 AF (0.1 cubic kilometers) in 2011 [30]. During the dry seasons, the withdrawal of large volumes of water for fracturing processes could significantly limit water availability for human consumption, crop, and livestock use [23]. Furthermore, since hydraulic fracturing has expanded to the drier southern and western parts of the state of Texas, the industry might have to adapt to those new conditions by reducing fresh water consumption and increasing water recycling and reuse [30].

11.2.2 HYDRAULIC FRACTURING AND LOCAL WATER QUALITY

Once hydraulic fracturing is complete, the pressure in the well is released by the removal of the pressure barriers such as the frac plugs [31]. Typically, two types of waste fluids, the flowback fluid and produced fluid, will be brought back to the surface from hydraulically fractured wells. The completion of hydraulic fracturing is accompanied with the quick flow back of hydraulic fluid mixed with brine (termed as "flowback" water) from the formation to the surface right before the well is placed into production for an average period of two to four weeks [10, 32]. The flowback will relieve the downhole pressure and allow gas migration to the surface. Once the well is placed into production, waste fluid (termed as "produced water" which includes subsequently returned hydraulic fluids and natural formation water) is continuously coproduced with gas over the lifetime of the well [33]. The amount and composition of wastewater generated by a particular well varies, however, greatly depending upon the geologic formation from which it originates, the extraction method utilized in the natural gas production process, and the chemicals (i.e., corrosion inhibitors and breakers) selected for the process [33–35]. The waste fluid (flowback as well as produced water) contains brine, fracturing fluid additives, hydrocarbons, and suspended and dissolved constituents from the shale formation and sometimes naturally occurring radioactive materials [10, 32–34]. The longer the fluid takes to return to the surface, the greater the concentration of formation materials will be found in the waste fluid [10, 36]. The water fluid is usually stored in onsite tanks or pits and is later treated either onsite or in another facility during the waste management process to reduce the toxicity of the fluid and minimize its environmental impacts.

Between 10 and 80% of the injected fracturing fluid volume may return to the surface as wastewater [36]. Both inorganic and organic constituents exist in flowback and produced water with in general inorganic components being much more extensive and prevalent than those of organic constituents [37]. The waste fluid generated from hydraulic fracturing wells is managed by deep well disposal, onsite treatment, reuse, or transportation offsite to treatment facilities followed by surface discharge [37]. The Pennsylvania Department of Environmental Protection (PADEP) requested that unconventional natural gas drillers voluntarily stop sending the wastewaters to Publicly-Owned Treatment Works (POTW) or Private Centralized Wastewater Treatment (CWT) within the commonwealth by May 19, 2011 [38], both of which might not be equipped to handle hydraulic fracturing wastewater [36]. In POTW of Pennsylvania, a typical treatment process includes a pretreatment to remove total suspended solids (TSS), followed by physical filtration, flocculation, aerobic digestion, and clarification before discharging into the surface water [36]. These processes are expected to remove organic compounds through degradation, but the removal of soluble, inorganic elements is less effective [36].

In commercial waste water treatment plants, Na_2SO_4 is added to first remove salts and metals as a solid precipitate prior to being treated by other processes [36]. The solids generated are then dried and hauled to residual waste landfills. While the commercial processes are expected to precipitate dissolved cations and filter solid elements, they are not expected to impact dissolved anions, such as chlorides as well as total dissolved solids (TDS) [36]. Limited research data have shown that prior to the voluntarily cessation of sending hydraulic fracturing waste to wastewater treatment plants, high levels of barium (Ba), strontium (Sr), bromides (Br), chlorides (Cl), TDS, and benzene were detected from the effluent discharge not only from POTW but also from CWT [36]. For instance, in Josephine Brine Treatment Inc., a commercially operated industrial wastewater treatment plant, Ba was detected from the effluent discharge with a mean concentration of 27.3 mg/L, 14 times EPA's maximum concentration limit (MCL) in drinking water and 4 times the derived drinking water minimal risk level (MRL) for intermediate and chronic exposures for adult men; 4.7 times the derived drinking water MRL for intermediate and chronic exposures for adult women; and 9 times the derived drinking water MRL for intermediate and chronic exposures for children [38]. The concentration of Ba in the effluent is also 1.3 and 6.7 times EPA criteria maximum concentration (CMC, 21 mg/L) and continuous concentration (CCC, 4.1 mg/L), respectively, the criteria set to protect aquatic health [38].

The mean concentration of Sr detected in the effluent of same facility was 2,980 mg/L, more than 740 times of EPA's recommended level in finished municipal drinking water of 4 mg/L, 43, 51 and 97 times the derived Sr drinking water MRL's for intermediate exposures for adult men, adult women, and children, respectively [38]. It is worth noting that wastewater treatment facility in Pennsylvania is required to report to PADEP for routine discharge of a toxic with a concentration above 100 µg/L (500 µg/L for nonroutine discharge); however, the review of documents reveals no evidence of notification by waste treatment to PADEP [38]. No criteria have been established by EPA for the concentration of Br in the drinking water to minimize forming of halogenated byproducts, the presence of which in high concentrations may be linked to the increase in cancerous diseases [39]. In the effluent of the Josephine Brine Treatment plant, a mean concentration of 1,070 mg/L for Br was detected, 10,700 times the concentration of 0.1 mg/L reported to be associated with adverse health concerns [39]. Again, as for Sr, a careful review of the archived documents revealed no notification of the discharge of high concentrations of bromine to PADEP [39].

The concentration of above mentioned analytes in the effluent of wastewater treatment facilities decreased significantly after the discharge of hydraulic fracturing wastewater into surface water was discontinued per PADEP's request, indicating the elevated concentrations of inorganic analytes found in the effluent prior to the voluntarily cease of discharge were largely attributable to the fracturing fluids [36], although more extensive investigations are still needed to confirm [36]. In 2011, EPA in conjunction with PADEP further requested that when considering the acceptance of fracking wastewater, the wastewater treatment plants need to document the chemicals used in hydraulic fracturing process that could reasonably be expected to be present in the wastewater and assess their potential impact on wastewater treatment and the receiving waters [40, 41]. Since then, the tighter regulation influenced several wastewater treatment plants to stop or reduce receiving unconventional natural gas fracturing fluid [36].

Fontenot et al. [42] reported elevated concentrations of arsenic, selenium, strontium, and TDS in some private water wells located near an active natural gas extraction site in Barnett Shale formation; however, it is important to recognize that there were also a number of private water wells in close proximity to natural gas wells that showed no elevated constituents. The spatial and temporal geochemical signatures of brines collected from shallow aquifers are compared with that from deeper shale formations in the Appalachian basin in the northeastern portion of Pennsylvania to assess the migration possibility of hydraulic fracturing fluid [43]. The elevated Br/Cl ratio (>0.001) and low Na/Cl ratio (<5) among other geochemical signatures found in a subset of shallow aquifers samples that lack geospatial association with the nearest shale gas wells were not distinguishable from the samples collected historically from the deeper Appalachian formations in the 1980s, the time period that no gas drilling activities were ongoing in the region [43]. These data delineate the possible natural mixing between the Appalachian brines and shallow groundwater through natural flow paths (i.e., fracture zones) that occurs over time and refute the overly broad claim that hydraulic fracturing accounts for the elevated groundwater salinity in all locations. On the other hand, these data suggest that great caution should therefore be taken prior to granting hydraulic fracturing or wastewater disposal injection activities in these areas because of the preexisting network of cross-formational pathways that connect to shallow groundwater specifically under high hydrodynamic pressure [43].

The organic constituents of produced water (i.e., organic acids and semivolatile organics) have also been studied [37]. These organic constituents in produced water are either attributable to the fracturing fluid additives or are from the release of natural organic compounds associated with formation which could comprise a significant portion of the organic matrix of the produced water. Benzene, toluene, ethylene, and xylene (aka BTEX) are commonly found in produced water [44]. Diesel fuel which introduces BTEX has been used as an additive to increase the efficiency in transporting proppants in the fracturing fluids [16]. BTEX are relatively mobile as well as toxic and/or confirmed carcinogens [45]. In 2004, EPA entered into a Memorandum of Agreement (MOA) with major service companies to voluntarily eliminate diesel fuel from hydraulic fracturing fluids that are injected directly into underground sources of drinking water (USDWs) for coal-bed methane (coal seam gas) production [16]. Similarly, the identification of trace amount of BTEX in well water near Miles, in western Queensland, Australia, led to the ban of the use of BTEX chemicals in fracturing fluids with the concern of adverse impacts on ground water [46]. While the ban or phase out of the use of diesel fuel in the hydraulic fracturing will reduce the introduction of BTEX into groundwater, BTEX also naturally exists in the gas and coal-bed deposition, the hydraulic fracturing process itself can therefore release significant amounts of BTEX into produced water even without using diesel fuel. It has been reported that a significant portion of wells drilled nationwide (56%) produced both oil and natural gas [47]. This is consistent with the report by Sirivedhin and associates that revealed that benzene-based compounds, particularly BTEX, were the dominant organic species found in produced water samples generated from oil/coalbed methane wells [37]. In addition to BTEX, dissolved organic acids from the formation such as monocarboxylic and dicarboxylic acid anions comprise the bulk of dissolved organic species found in produced water.

The remaining portion of fracturing fluid that is left far beneath groundwater levels may induce the opening of long fissures over time which will allow fracturing fluid or natural gas to travel upward to reach groundwater and thus reducing its quality. It has been assumed that in Marcellus shale, hydraulic fractures are confined vertically and the hydraulic fracturing process is conducted thousands of feet below the deepest aquifers suitable for drinking water [19]. Recent findings might shift this paradigm. Osborn and colleagues demonstrate that in active gas-extraction areas with one or more gas wells within 1 km (3,280 feet) apart, both average and maximum methane concentrations in drinking-water wells increase with proximity to the nearest gas well (19.2 and 64 mg/L, resp.) in northeastern Pennsylvania [48]; in contrast, in neighboring nonextraction sites within similar geologic formations methane concentrations only averaged 1.1 mg/L and the farther away from natural gas development, the lower combustible gas concentrations were found in water wells [16, 23]. However, the study surveyed a relative small number of nonrandomized wells, of which several of the contaminated water wells reported in the study were from a region that had aquifer contamination in the past that might be associated with casing leaks or inadequate cementing of orphaned gas wells rather than hydraulic fracturing. In addition, Osborn's study did not include baseline measurements of levels of methane in aquifers prior to fracturing [49–51], without of which any definitive conclusions are questionable. Therefore, caution is needed to interpret and extrapolate Osborn's results systematically and long-term, coordinated sampling and monitoring procedures are required for future studies.

11.2.3 WASTEWATER MANAGEMENT CONCERNS

Public concern has been voiced about the potential leakage of fracturing wastewater into other water bodies, if not probably stored, treated, or disposed. Gross and associates have demonstrated that surface spills or leakage into the shallow water formations is a critical event and could account for most water quality issues associated with hydraulic fracturing [52]. Chemicals can potentially leach into groundwater through failures in the lining of ponds or containment systems most of which are constructed near the well sites to temporarily hold flowback/produced water [52–54]. Between 2009 and 2010, of the 4,000 permitted oil and natural gas wells in Marcellus Shale in Pennsylvania, there were 630 reported environmental health and safety violations of which half were associated with leaks and spills of the flowback/produced fluids [55]. In Weld County, Colorado, groundwater is the main source of water supply for local and commercial use. Some locations of Weld County have very shallow depth to water table with higher opportunities for groundwater contamination, yet Weld County has the highest density of wells used by hydraulic fracturing for both natural gas and crude oil in the United States. In Weld County, tank battery systems (for storing produced water and crude oil in various stages of separation) and production facilities (sources of hydrocarbons in the refining process) were found at most well sites which could contribute to leaks and spills [52]. BTEX for example, can pass through soil into the groundwater after spills. Gross assessed the profiles of BTEX concentrations over time by repeated sampling of groundwater on/near multiple spill sites either prior to or shortly after remediation began. In total, 90% of groundwater samples collected contained benzene and 30% of samples contained toluene concentrations above their MCLs. Although there was a delay between the reported surface spill date and the first water sampling date (BTEX is volatile; it is possible that the initial BTEX concentration shortly after spills could be even higher), nevertheless the mean concentration of benzene and toluene from groundwater samples collected from inside the excavation sites prior to the remediation was 280 and 2.2 times that of respective MCLs [52]. While these data indicate that benzene and toluene are of greater concern when considering BTEX groundwater concentrations from surface spills, the results in the report should be interpreted with caution. The distribution of BTEX concentrations collected at spill sites is highly skewed with the median values much lower than the estimated means, in some cases several hundredfold lower. For example, the median value for toluene does not exceed its respective MCL [52]. In addition, the study fails to differentiate the contamination sources of BTEX which could come from onsite oil spills, natural gas-related produced water spills, or a combination of the two, although produced water from gas production could have higher contents of BTEX than water from oil production [44].

Flowback/produced water reuse for hydraulic fracturing is another option chosen by an increasing number of oil and gas companies as it reduces wastewater generated, fresh water required, and wastewater management costs. Currently, most of the flowback water from Marcellus

Shale in Pennsylvania is recycled and reused in future hydraulic fracturing processes, while industrial treatment and discharge into surface water declined to only 3% [56]. The fracturing fluid reuse option however is not without limitations. Not all produced water from the hydraulic fracturing is suitable for reuse. The chemical signatures as well as the concentrations of TDS, TSS, and brines in produced water largely depend on the nature of the formation [33]. Highly soluble TDS could be difficult and expensive to remove from wastewater, and their combination with various other contaminants necessitates multiple treatment technologies in sequence [56]. For instance, the quality of produced water from Haynesville Shale is considered less attractive for reuse potentials as it contains high levels of TDS, chlorides, and TSS and has high scaling tendency (high calcium and high magnesium). In contrast, produced water from Fayetteville Shale is considered to have excellent potential for reuse due to low concentrations of chlorides, TDS, and low scaling tendency [33]. In Marcellus Shale, however, only TSS, but not the salts in produced water, is filtered prior to reuse. Therefore, the concentrations of the remaining components (i.e., TDS/brines, scaling components) if beyond accepted range need to be diluted with substantial amounts of fresh water before the filtered produced water can be directly reused [33]. The amount of energy required to treat the fluids, the amount of air pollution generated, the amount of solid waste that will be disposed of in landfills, and the cost of logistics are among other factors that will determine the feasibility of the produced water reuse option. In addition, some shale formations tend to either "trap" or are considered as highly desiccated which render significant insufficient amounts of flowback fluid back to the surface [33].

Several other mechanisms for wastewater management have also been applied. Wastewater can be injected underground into Class II Wells [57, 58]. Class II disposal wells which account for approximately 20% of 144,000 Class II Wells in the United States can only be used to dispose fluids associated with oil and gas production. While UIC requirements do not apply to hydraulic fracturing drilling operation (with exception when diesel fuel is used), the underground injection of wastewater generated during oil and gas production (including hydraulic fracturing) requires an Underground Injection (UIC) permit under the SDWA [58]. In many regions of the United States, underground injection is the most common method of disposing fluids or other substances generated from shale gas extraction operations [59]. For instance, approximately 98% of all brine is disposed of by injection back into brine-bearing or depleted oil and gas formations deep below the earth's surface in Ohio [60].

While wastewater injection into Class II Wells may be an ideal solution for some states perhaps due to the presence of suitable geological formation in the area as well as the availability of sufficient wells, this option may be problematic for other states that have already reached their maximum injection capacity or are unable to use this option due to unsuitable geology [43, 61]. The existence of cross-formational pathways allowing deep saline water to migrate upward into shallow, fresher aquifers has been documented in Appalachian Basin as well as in other areas across the nation [43]. Furthermore, in Pennsylvania, more than 180,000 wells had been drilled prior to any requirement for documenting the locations. The location of many wells is unknown, while others have been improperly abandoned [62]. The existence of these wells might increase the chance of injected waste fluids escaping the injection formation through the connecting fractures and transport to higher aquifer, although this issue is still up for debate [63, 64]. As a result, wastewater in Pennsylvania has been transported to nearby states for treatment which presents concerns of increased possibility of leaks or spills during transportation [32]. Clearly, more data are required in order to further evaluate the movement of contaminations along pathways either from wellbores or from deep formations to overlying groundwater.

The wastewater can also be treated and reused in irrigation, for unpaved road dust control, or even roads deicing, although the portion of the waster fluid used for these purposes is very small, having received a high degree of criticism and is discouraged [35, 56, 65]. In these cases, the wastewater is either treated or mixed with large volumes of fresh water to lower its TDS and other constituents to acceptable ranges [33]. These solutions inevitably require additional water withdrawals, increased onsite storage capacity, increased cost of transportation, and requires more resources and chemicals for treatment purposes. Moreover, the use of wastewater for crop irrigation or unpaved road dust control may pose additional health threat due to the unknown toxicity of many individual components attributable to the trade secret nature of many constitutes used by hydraulic fracturing companies [65]. While no published data are currently available for assessing the potential environmental and human and animal health impacts of treated produced water used for abovementioned purposes, we have learned a very expensive lesson in the Times Beach site in Missouri at the expense of the disincorporation of the city [66]. The site was sprayed with waste oil in early 1970s for unpaved dust control; in 1982, EPA later found the oil to be contaminated with dioxins which are highly toxic and can cause severe reproductive and developmental problems.

11.3 CONCLUSION

While hydraulic fracturing may present an economic advantage to the United States by transitioning the country to an energy independent state, there are several environmental concerns associated with the process that have not been properly addressed.

Primarily, a main concern of the public and environmentalists pertaining to hydraulic fracturing is governmental leniency in its regulation. In the 2005 Energy Policy Act, Congress revised the SDWA definition of "underground injection" to specifically exclude the "underground injection of fluids or propping agents (other than diesel fuels) pursuant to hydraulic fracturing operations related to oil, gas, or geothermal production activities" from UIC regulation (SDWA Section 1421(d)(1)(B)) [67]. Between 2005 and 2009, the 14 leading oil and gas service companies used 780 million gallons of chemical products in fracturing fluids [68]. The concentration and composition of the fluid used in hydraulic fracturing vary with the nature of the formation. Although some of these chemicals may be harmless, others are not well investigated and may be hazardous to human health and the environment. To allow for a competitive market in the field of hydraulic fracturing, under current regulation, oil and gas companies are not required to disclose the identity of the chemicals in their fracturing fluids other than under the Emergency Planning and Community Right-to-Know Act (EPCRA), under which owners or operators of facilities where certain hazardous hydraulic fracturing chemicals are present above certain thresholds may have to comply with emergency planning requirements, emergency release notification obligations, and hazardous chemical storage reporting requirements [69]. While disposal management of fracturing wastes is regulated, Provisions of the Resource Conservation and Recovery Act (RCRA) exempt drilling fluids, produced water, and other wastes associated with the exploration, development, or production of crude oil, natural gas, or geothermal energy from regulation as hazardous wastes under Subtitle C of RCRA. As a consequence, instead of being disposed into Class I Wells which are designated for both hazardous and nonhazardous waste, wastewater is injected into Class II Wells which are far less regulated than Class I [69]. In addition, appropriate documentation/registration/reporting systems for hydraulic fracturing related activities are loosely enforced and the implementation status of current law regulation is less than satisfactory. In August 2013, the United States Government Accountability Office (GAO) pointed out that current Bureau of Land Management's (BLM) environmental inspection prioritization process may miss oil and gas wells that could pose the greatest environmental risk [70]. BLM manages onshore federal oil and gas resources and ensure that oil and gas operations on federal lands are prudently conducted in a manner that ensures protection of the surface and subsurface environment. However, approximately 41% of the 60,330 federal oil and gas wells including those used for wastewater disposal found no record of ever having received an environmental inspection between 2007 and 2012 [70]. Similarly, in the case of hydraulic fracturing fluid spills, the majority of reports did not include information specifying the volume of produced water that was spilled [52].

Hydraulic fracturing is a major investment of several countries globally. Canada, South Africa, Germany, United Kingdom, Russia, and China all use fracturing techniques to increase their natural gas production. The choice to continuously conduct hydraulic fracturing is currently under debate in the United Kingdom due to public concerns of its potential environmental impacts [71]. France and Bulgaria on the other hand have banned the use of hydraulic fracturing for gas extraction because of environmental concerns [72, 73]. In United States, if hydraulic fracturing results in the release of hazardous substances at or under the surface in a manner that may endanger public health or the environment [69], all potentially responsible parties could face liability under CERCLA for cleanup costs, natural resource damages, and the costs of federal public health studies. However, federal regulations on fracturing overall have not been stringent enough. Proper documentation/reporting systems for wastewater discharge and spills need to be enforced at the federal, state, and industrial level and UIC requirements under SDWA should be extended to hydraulic fracturing operations regardless if diesel fuel is used as a fracturing fluid (or a component of a fracturing fluid) or not. Furthermore, federal laws mandating hydraulic companies to disclose fracturing fluid composition and concentration not only to federal and state regulatory agencies but also to health care professionals would encourage this practice. Only the full disclosure of fracturing chemicals will allow future research to fill the knowledge gaps for a better understanding of the impacts of hydraulic fracturing on human health and environment [19, 20] as well as to determine if any further regulations or the improvement of technology itself are needed.

REFERENCES

1. U.S. Energy Information Administration, "U.S. Energy Facts Explained," 2013, http://www.eia.gov/energyexplained/index.cfm?page=us_energy_home.
2. U.S. Department of State, "Fifth climate action report to the UN framework convention on climate change," 2010, http://www.state.gov/e/oes/rls/rpts/car5/.
3. U.S. Energy Information Administration, "Monthly Energy Review DOE/EIA, 0035 (2013/11)," 2013, http://www.eia.gov/totalenergy/data/monthly/archive/00351311. pdf.
4. D. Weisser, "A guide to life-cycle greenhouse gas (GHG) emissions from electric supply technologies," Energy, vol. 32, no. 9, pp. 1543–1559, 2006.
5. K. Ritter, A. Emmert, M. Lev-On, and T. Shires, "Understanding greenhouse gas emissions from unconventional natural gas production," in Proceedings of the 20th International Emissions Inventory Conference, Tampa, Fla, USA, 2012, http://www. epa.gov/ttnchie1/conference/ei20/session3/kritter.pdf.
6. M. Ratner and M. Tiemann, "An overview of unconventional oil and natural gas: resources and federal actions," Congressional Services Report, 2013, https://www. fas.org/sgp/crs/misc/R43148.pdf.
7. U.S. Energy Information Administration, "Natural Gas Year-in-Review," 2012, http://www.eia.gov/naturalgas/review/archive/2011/.
8. U.S. Government Accountability Office, "Oil and Gas Information on Shale Resources, Development, and Environmental and Public Health Risks," 2012, http:// www.gao.gov/products/GAO-12-732.
9. L. D. Helms, "Horizontal drilling," North Dakota Department of Mineral Resources Newsletter, vol. 35, no. 1, pp. 1–3, 2008.

10. L. Haluszczak, A. W. Rose, and L. Kump, "Geochemical evaluation of flowback brine from Marcellus gas wells in Pennsylvania, USA," Applied Geochemistry, vol. 28, pp. 55–61, 2013.

11. R. E. Jackson, A. W. Gorody, B. Mayer, et al., "Groundwater protection and unconventional gas extraction: the critical need for field-based hydrogeological research," Ground Water, vol. 51, no. 4, pp. 488–510, 2013.

12. FracFocus, "Chemical use in hydraulic fracturing," Chemical Disclosure Registry, 2013, http://fracfocus.org/water-protection/drilling-usage.

13. R. McCurdy, "High rate hydraulic fracturing additives in nonmarcellus unconventional shales," EPA Hydraulic Fracturing Workshop, 2011, http://www2.epa.gov/sites/production/files/documents/highratehfinnon-marcellusunconventionalshale.pdf.

14. J. D. Arthur, B. Bohm, B. J. Coughlin, and M. Layne, "Evaluating the environmental implications of hydraulic fracturing in shale gas reservoirs," ALL Consulting, 2008, http://www.all-llc.com/publicdownloads/ArthurHydrFracPaperFINAL.pdf.

15. K. K. Mohanty, A. Gaurav, and M. Gu, "Improvement of Fracturing for Gas Shales, Final report to Research Partnership to Secure Energy for America," 2012, http://www.rpsea.org/projects/07122-38/.

16. U.S. Environmental Protection Agency, "Evaluation of Impacts to Underground Sources of Drinking Water by Hydraulic Fracturing of Coalbed Methane Reservoirs Study," 2004, http://water.epa.gov/type/groundwater/uic/class2/hydraulicfracturing/wells_coalbedmethanestudy.cfm.

17. P. Kaufman, G. S. Penny, and J. Paktinat, "SPE 119900 critical evaluations of additives used in shale slickwater fracs," in Proceedings of the SPE Shale Gas Production Conference, Irving, Tex, USA, November 2008, http://www.flotekind.com/Assets/SPE-119900-Critical-Evaluations.pdf.

18. M. McFeeley, "State hydraulic fracturing disclosure rules and enforcement: a comparison," Natural Resources Defense Council, 2012, http://www.nrdc.org/energy/files/Fracking-Disclosure-IB.pdf.

19. A. L. Maule, C. M. Makey, E. B. Benson, I. J. Burrows, and M. K. Scammell, "Disclosure of hydraulic fracturing fluid chemical additives: analysis of regulations," New Solution, vol. 23, no. 1, pp. 167–187, 2013.

20. M. A. Rafferty and E. Limonik, "Is shale gas drilling an energy solution or public health crisis?" Public Health Nursing, vol. 30, no. 5, pp. 454–462, 2013.

21. FracFocus, "Hydraulic fracturing: the process," Chemical Disclosure Registry, 2013, http://fracfocus.org/hydraulic-fracturing-how-it-works/hydraulic-fracturing-process.

22. U.S. Department of Energy, "Modern Shale Gas Development in the United States: A Primer," 2009, http://www.netl.doe.gov/technologies/oil-gas/publications/EPreports/Shale_Gas_Primer_2009.pdf.

23. H. Cooley and K. Donnelly, "Hydraulic Fracturing and Water Resources: Separating the Frack from the Fiction," Pacific Institute, Oakland, CA, USA, 2012, http://www.pacinst.org/wp-content/uploads/2013/02/full_report35.pdf.

24. U.S. Environmental Protection Agency, "Plan to Study the Potential Impacts of Hydraulic Fracturing on Drinking Water Resources," 2011, http://water.epa.gov/

type/groundwater/uic/class2/hydraulicfracturing/upload/hf_study_plan_110211_final_508.pdf.

25. Chesapeaker Energy, "Water Use in Eagle Ford Deep Shale Exploration," 2012, http://www.chk.com/media/educational-library/fact-sheets/eagleford/eagleford_water_use_fact_sheet.pdf.

26. E. G. Johnson and L. A. Johnson, "Hydraulic fracture water usage in northeast British Columbia: locations, volumes and trends," in Geoscience Reports, pp. 41–63, British Columbia Ministry of Energy and Mines, British Columbia, Canada, 2012.

27. J. F. Kenny, N. L. Barber, S. S. Hutson, et al., "Estimated use of water in the United States in 2005," U.S. Geological Survey Circular 1344, 2009, http://pubs.usgs.gov/circ/1344/pdf/c1344.pdf.

28. B. Poulson, "Weld County, Colorado: Ground Zero in the Anti-fracking Battle, Forbes," 2013, http://www.forbes.com/sites/realspin/2013/12/04/weld-county-colorado-ground-zero-in-the-anti-fracking-battle/.

29. S. Goldenberg, "A Texan Tragedy: Ample Oil, No Water, The Guardian," 2013, http://www.theguardian.com/environment/2013/aug/11/texas-tragedy-ample-oil-no-water.

30. J. P. Nicot, R. C. Reedy, R. A. Costley, and Y. Huang, "Oil & Gas Water Use in Texas: Update to the 2011 Mining Water Use Report," Bureau of Economic Geology, Jackson School of Geosciences and The University of Texas at Austin, 2012, http://www.twdb.state.tx.us/publications/reports/contracted_reports/doc/0904830939_2012Update_MiningWaterUse.pdf.

31. Cabot Oil and Gas Corporation, Exploring the Hydraulic Fracturing Process, 2013, http://www.cabotog.com/wp-content/uploads/2013/06/HydraulicFracturing-II.pdf.

32. R. D. Vidic, S. L. Brantley, J. M. Vandenbossche, D. Yoxtheimer, and J. D. Abad, "Impact of shale gas development on regional water quality," Science, vol. 340, no. 6134, Article ID 1235009, 2013.

33. M. Mantell, "Produced water reuse and recycling challenges and opportunities across major shale plays," in Proceedings of the EPA Hydraulic Fracturing Study Technical Workshop #4 Water Resources Management, March 2011, http://www2.epa.gov/sites/production/files/documents/09_Mantell_-_Reuse_508.pdf.

34. A. Murali Mohan, A. Hartsock, K. J. Bibby, et al., "Microbial community changes in hydraulic fracturing fluids and produced water from shale gas extraction," Environmental Science & Technology, vol. 47, no. 22, pp. 13141–13150, 2013.

35. J. A. Veil, M. G. Pruder, D. Elcock, and R. J. Redweik Jr., A White Paper Describing Produced Water From Production of Crude Oil, Natural Gas, and Coal Bed Methane, Prepared for Department of Energy, National Energy Technology Laboratory, 2004, http://www.circleofblue.org/waternews/wp-content/uploads/2010/08/prodwaterpaper1.pdf.

36. K. J. Ferrar, D. R. Michanowicz, C. Christen, et al., "Assessment of effluent contaminants from three facilities discharging marcellus shale wastewater to surface waters in Pennsylvania," Environmental Science & Technology, vol. 47, no. 7, pp. 3472–3481, 2013.

37. T. Sirivedhin and L. Dallbauman, "Organic matrix in produced water from the Osage-Skiatook Petroleum Environmental Research site, Osage county, Oklahoma," Chemosphere, vol. 57, no. 6, pp. 463–469, 2004.

38. C. D. Volz, "Written testimony before the senate committee on environment and public works and its subcommittee on water and wildlife, joint hearing," Natural Gas Drilling, Public Health and Environmental Impacts, 2011, http://www.epw.senate.gov/public/index.cfm?FuseAction=Files.View&FileStore_id=57d1bfd4-9237-488e-999f-4e1e71f72e52.

39. L. T. Stayner, M. Pedersen, E. Patelarou, et al., "Exposure to brominated trihalomethanes in water during pregnancy and micronuclei frequency in maternal and cord blood lymphocytes," Environmental Health Perspectives, 2013. View at Publisher · View at Google Scholar

40. U.S. Environmental Protection Agency, Letters from EPA to 14 Publicly Owned Treatment Works (POTWs) in Pennsylvania Stating that Drilling Waste May Be Considered a Substantial Change, Key Documents About Mid-Atlantic Oil and Gas Extraction, 2011, http://www.epa.gov/region03/marcellus_shale/pdf/potw7-13-11/altoona.pdf.

41. U.S. Environmental Protection Agency, Letters to Pennsylvania Publicly Owned Treatment Works (POTWs) regarding acception of oil and gas waste, Key Documents About Mid-Atlantic Oil and Gas Extraction, 2013, http://www.epa.gov/region03/marcellus_shale/pdf/oandg-wpretreatment.pdf.

42. B. E. Fontenot, L. R. Hunt, Z. L. Hildenbrand, et al., "An evaluation of water quality in private drinking water wells near natural gas extraction sites in the barnett shale formation," Environmental Science & Technology, vol. 47, no. 17, pp. 10032–10040, 2013.

43. N. R. Warner, R. B. Jackson, T. H. Darrah, et al., "Geochemical evidence for possible natural migration of Marcellus Formation brine to shallow aquifers in Pennsylvania," Proceedings of the National Academy of Sciences, vol. 109, no. 30, pp. 11961–11966, 2012.

44. K. Guerra, K. Dahm, and S. Dundorf, "Oil and gas produced water management and beneficial use in the Western United States," U.S. Department of the Interior, Science and Technology Program Report No. 157, 2011, http://www.usbr.gov/research/AWT/reportpdfs/report157.pdf.

45. Y. K. Kunukcu, "In situ bioremediation of groundwater contaminated with petroleum constituents using oxygen release compounds (ORCs)," Journal of Environmental Science and Health A: Toxic/Hazardous Substances and Environmental Engineering, vol. 42, no. 7, pp. 839–845, 2007.

46. S. Robertson, BTEX Ban Enforced through Legislation, Queensland Government, 2010, http://statements.qld.gov.au/Statement/Id/71854.

47. U.S. Energy Information Administration, Rethinking Rig Count As a Predictor of Natural Gas Production, 2013, http://www.eia.gov/todayinenergy/detail.cfm?id=13551.

48. S. G. Osborn, A. Vengosh, N. R. Warner, and R. B. Jackson, "Methane contamination of drinking water accompanying gas-well drilling and hydraulic fracturing," Proceedings of the National Academy of Sciences of the United States of America, vol. 108, no. 20, pp. 8172–8176, 2011.

49. S. C. Schon, "Hydraulic fracturing not responsible for methane migration," Proceedings of the National Academy of Sciences of the United States of America, vol. 108, no. 37, p. E664, 2011.

50. T. Saba and M. Orzechowski, "Lack of data to support a relationship between methane contamination of drinking water wells and hydraulic fracturing," Proceedings of the National Academy of Sciences of the United States of America, vol. 108, no. 37, p. E663, 2011.

51. R. J. Davies, "Methane contamination of drinking water caused by hydraulic fracturing remains unproven," Proceedings of the National Academy of Sciences of the United States of America, vol. 108, no. 43, p. E871, 2011.

52. S. A. Gross, H. J. Avens, A. M. Banducci, et al., "Analysis of BTEX groundwater concentrations from surface spills associated with hydraulic fracturing operations," Journal of Air & Waste Management Association, vol. 63, no. 4, pp. 424–432, 2013.

53. M. Metzger, "Hydrofracturing and the environment," Water Quality Products, vol. 16, no. 11, pp. 14–15, 2011.

54. K. B. Gregory, R. D. Vidic, and D. A. Dzombak, "Water management challenges associated with the production of shale gas by hydraulic fracturing," Elements, vol. 7, no. 3, pp. 181–186, 2011.

55. D. J. Rozell and S. J. Reaven, "Water pollution risk associated with natural gas extraction from the marcellus shale," Risk Analysis, vol. 32, no. 8, pp. 1382–1393, 2012.

56. B. G. Rahm, J. T. Bates, L. R. Bertoia, et al., "Wastewater management and Marcellus Shale gas development: trends, drivers, and planning implications," Journal of Environmental Management, vol. 120, pp. 105–113, 2013.

57. U.S. Environmental Protection Agency, Permitting Guidance for Oil and Gas Hydraulic Fracturing Activities Using Diesel Fuels. Draft—Underground Injection Program Guidance #84, 2012, http://water.epa.gov/type/groundwater/uic/class2/hydraulicfracturing/upload/hfdieselfuelsguidance508.pdf.

58. U.S. Environmental Protection Agency, EPA Class II Wells—Oil and Gas Related Injection Wells (Class II), 2012, http://water.epa.gov/type/groundwater/uic/class2/.

59. C. E. Clark and J. A. Veil, Produced Water Volumes and Management Practices in the United States, Environmental Science Division, Argonne National Laboratory, 2009, http://www.ipd.anl.gov/anlpubs/2009/07/64622.pdf.

60. Ohio Department of Natural Resources, Underground Injection Control (UIC), 2013, http://oilandgas.ohiodnr.gov/industry/underground-injection-control.

61. W. L. Ellsworth, "Injection-induced earthquakes," Science, vol. 341, no. 6142, Article ID 1225942, 2013.

62. T. Myers, "Potential Contaminant Pathways from Hydraulically Fractured Shale to Aquifers," Ground Water, vol. 50, no. 6, pp. 872–882, 2012.

63. H. A. Cohen, T. Parratt, and C. B. Andrews, "Potential contaminant pathways from hydraulically fractured shale to aquifers," Ground Water, vol. 51, no. 3, pp. 317–321, 2013.

64. T. Myers, "Potential Contaminant Pathways from Hydraulically Fractured Shale to Aquifers," Ground Water, vol. 50, no. 6, pp. 826–830, 2012.

65. L. Shariq, "Uncertainties associated with the reuse of treated hydraulic fracturing wastewater for crop irrigation," Environmental Science & Technology, vol. 47, no. 6, pp. 2435–2436, 2013.

66. M. Gough, "Human exposures from dioxin in soil—a meeting report," Journal of Toxicology and Environmental Health, vol. 32, no. 2, pp. 205–235, 1991. View at Scopus

67. U.S. Environmental Protection Agency, Regulation of Hydraulic Fracturing under the Safe Drinking Water Act, 2012, http://water.epa.gov/type/groundwater/uic/class2/hydraulicfracturing/wells_hydroreg.cfm.

68. U.S. House of Representatives Committee on Energy and Commerce Minority Staff, Chemical Used in Hydraulic Fracturing, 2011, http://democrats.energy-commerce.house.gov/sites/default/files/documents/Hydraulic-Fracturing-Chemicals-2011-4-18.pdf.

69. A. Vann, B. J. Murrill, and M. Tiemann, Hydraulic Fracturing: Selected Legal Issues, Congressional Research Service, 2013, http://www.fas.org/sgp/crs/misc/R43152.pdf.

70. U.S. Government Accountability Office, Oil and Gas Development: BLM Needs Better Data to Track Permit Processing Times and Prioritize Inspections, 2013, http://www.gao.gov/assets/660/657176.pdf.

71. R. Jaspal and B. Nerlich, "Fracking in the UK press: threat dynamics in an unfolding debate," Public Understanding of Science, 2013. View at Publisher · View at Google Scholar

72. D. Jolly, France Upholds Ban on Hydraulic Fracturing, The New York Times, 2013, http://www.nytimes.com/2013/10/12/business/international/france-upholds-fracking-ban.html.

73. M. Bran, Bulgaria Becomes Second State to Impose Ban on Shale-Gas Exploration, The Guardian, 2012, http://www.theguardian.com/world/2012/feb/14/bulgaria-bans-shale-gas-exploration.

Author Notes

CHAPTER 1

Acknowledgment
The work was performed with financial support from the University of Texas Energy Institute. We thank the database provider, IHS (http://www.ihs.com), for access to the Enerdeq database. We also gratefully acknowledge the oil and gas companies that are members of the Barnett Shale Water Conservation and Management Committee, Groundwater Conservation Districts, River Authorities, and Water Districts for their time and for providing data. Chris Parker edited the manuscript. Publication authorized by the Director, Bureau of Economic Geology, The University of Texas at Austin.

CHAPTER 3

Acknowledgment
This work was supported by the U.S. Department of Energy, National Energy Technology Laboratory, Grant DE-FE0000975. The views and opinions of the authors expressed herein do not necessarily state or reflect those of the United States Government or any agency thereof.

CHAPTER 4

Acknowledgments
We thank Will Wheeler at the US EPA for data on NPDES-permitted facilities in Pennsylvania and New York, the Bureau of Topographic and Geologic Survey of the PADCNR for data on well completions, and Stefan Staubli. We gratefully acknowledge financial support from the Alfred P. Sloan Foundation.

CHAPTER 5

Acknowledgments
Funding for this study was provided by the Nicholas School of the Environment and the Center on Global Change at Duke University and by funding to the Nicholas School from Fred and Alice Stanback. Field sampling activities were funded by Shirley Community Development Corporation and Faulkner County, Arkansas. Any use of trade, firm, or product names is for descriptive purposes only and does not imply endorsement by the US Government.

CHAPTER 8

Acknowledgment
As part of the National Energy Technology Laboratory's Regional University Alliance (NETL–RUA), a collaborative initiative of the NETL, this study was performed under Task Release TR 131, Project Activity 4.605.920.009.812.

CHAPTER 11

Conflict of Interests
The authors declare that there is no conflict of interests regarding the publication of this paper.

Author Contributions
Dr. Chen and Dr. Al-Wadei contributed equally.

Index